Mars Exploration - a Step Forward

Edited by Giuseppe Pezzella and Antonio Viviani

Published in London, United Kingdom

IntechOpen

Supporting open minds since 2005

Mars Exploration - a Step Forward
http://dx.doi.org/10.5772/intechopen.78933
Edited by Giuseppe Pezzella and Antonio Viviani

Contributors
Ye Lu, Carole Tafforin, Radvan Bahbouh, Rafael Bardera, Suthyvann Sor Mendi, Adelaida García-Magariño, Haoting Liu, Jim Pass, Bryan Palaszewski, Andoni G. Moral, Guillermo Lopez-Reyes, Giuseppe Pezzella, Antonio Viviani

First published in London, United Kingdom, 2020 by IntechOpen
IntechOpen is the global imprint of INTECHOPEN LIMITED, registered in England and Wales, registration number: 11086078, 7th floor, 10 Lower Thames Street, London, EC3R 6AF, United Kingdom
Printed in Croatia

British Library Cataloguing-in-Publication Data
A catalogue record for this book is available from the British Library

Additional hard and PDF copies can be obtained from orders@intechopen.com

Mars Exploration - a Step Forward
Edited by Giuseppe Pezzella and Antonio Viviani
p. cm.
Print ISBN 978-1-83962-362-2
Online ISBN 978-1-83962-363-9
eBook (PDF) ISBN 978-1-83962-364-6

For EU product safety concerns:
IN TECH d.o.o., Prolaz Marije Krucifikse Kozulić 3, 51000 Rijeka, Croatia, info@intechopen.com or visit our website at intechopen.com.

Meet the editors

Prof. Giuseppe Pezzella was born on November 23, 1972, in Naples, Italy. He is a lecturer in Aerodynamics and Aerothermodynamics at the Department of Engineering of the University of Campania, Italy. He received an MSc in Aeronautical Engineering in 1999 from the University of Naples "Federico II." Prof. Pezzella obtained his PhD in Aerospace Engineering from the same university in 2003. He has been a research engineer in the Physics of Fluids Department at CIRA since 2005, after a period spent at University of Naples 'Federico II" in the frame of post-doc activities. He is a specialized analyst in the fields of vehicle aerodynamics, aerothermodynamics, and aeroshape design. He is involved in national (e.g., PRORA) and international projects (e.g., EXPERT, HIGH-LIFT, FLPP-IXV, RASTAS-SPEAR, HEXAFLY, LAPCAT-II, SCRAMSPACE, HEXAFLY-INT). Prof. Pezzella is author or co-author of about 100 publications among archive journals, conference papers, and edited books, and a reviewer of several international journals.

Antonio Viviani was born in Salerno, Italy. He has a graduate degree in Aeronautical Engineering and PhD in Aerospace Engineering from the University of Naples, Italy. He is a full professor in the Engineering Department of the University of Campania, Italy. He is a member of several organizations including the American Institute of Aeronautics and Astronautics (AIAA); American Association for Advancement of Sciences; European Low Gravity Research Association (ELGRA); International Academy of Astronautics (IAA). He also serves on many committees of: International Astronautical Federation (IAF), International Academy of Astronautics (IAA), International Council of Aeronautical Sciences (ICAS). Prof. Viviani has held several positions including vice president of ELGRA (1999-2003); vice chairman of the Committee of IAF Microgravity Sciences and Processes (2003-2006); chairman of the Committee of IAF Microgravity Sciences and Processes (2006-2012); now, IAA board of trustees member, and IAA regional secretary for Italy; representative for Italy in the ICAS member societies council; co-editor of *Acta Astronautica*, ed. board *Int. J. Aerospace Engineering*. He was co-investigator of the experiment "Onset of Oscillatory Marangoni Flows," Spacelab D2 - Space-Shuttle Columbia STS-55 (1993), and principal investigator of the experiments "Bubble Behavior under Low Gravity," Mission Spacelab IML-2 - Space-Shuttle Columbia, STS-65 (1994); and "Non-linear Surface Tension Driven Bubble Migration," Mission Spacelab LMS – Space Shuttle Columbia STS-78 (1996).

Contents

Preface

Mars exploration plays a fundamental role in space programs worldwide. More than 50 years after the Mariner 4 spacecraft flyby mission on 15 July 1965, the red planet still represents the next frontier of extra-terrestrial space exploration programs.

To date, several space agencies all over the world have sent spacecraft to Mars, but only two countries have accomplished successful landings. The Soviet Union landed on Mars for the first time in history with the Mars 3 and Mars 6 landings in 1971 and 1973, respectively. Since then, there have been eight successful United States Mars landings, namely Viking 1 and Viking 2 (both 1976), Pathfinder (1997), Spirit and Opportunity (both 2004), Phoenix (2008), Curiosity (2012), and InSight (2018).

Other countries have sent or aim to send spacecraft towards the red planet in the near future, including Europe, Japan, United Arabs Emirates, India, and China. For instance, there are three missions leaving for Mars within ten days at the end of July 2020. They are, in the order of launch, Hope of Emirates (July 20), Tianwen-1 of China (July 23), and Mars 2020 of Nasa (July 30).

However, sending a spacecraft to Mars demands huge efforts from both scientific and engineering points of view. Several issues must be addressed, including space probe design, launch, interplanetary journey, approach, entry and descent flights, and landing when a rover is present.

The next goal is to send human missions to Mars. Proposals for crewed exploration have been made throughout the history of space exploration; this has been an aspiration since the earliest days of space science.

This book contains three sections covering Missions to Mars, In Situ Operations, and Human - Rated Missions. Each section contains valuable contributions focusing on several mission design issues. Topics covered include discussion of psychological effects related with human-rated missions. The information contained herein will allow for the development of safe and efficient exploration missions to Mars.

In this framework, the aim of the book is to support industries, research centers, and space agencies in their own design and development of next-generation missions to Mars. Therefore, this book is recommended for both students and research engineers involved in all design phases typical for exploration missions to Mars.

Giuseppe Pezzella and Antonio Viviani
Professor,
Department of Engineering,
University of Campania "L. Vanvitelli",
Aversa, Italy

Section 1

Missions to Mars

Chapter 1

Introductory Chapter: Mars Exploration - A Story Fifty Years Long

Giuseppe Pezzella and Antonio Viviani

1. Introduction

Mars has been a goal of exploration programs of the most important space agencies all over the world for decades. It is, in fact, the most investigated celestial body of the Solar System. Mars robotic exploration began in the 1960s of the twentieth century by means of several space probes sent by the United States (US) and the Soviet Union (USSR). In the recent past, also European, Japanese, and Indian spacecrafts reached Mars; while other countries, such as China and the United Arab Emirates, aim to send spacecraft toward the red planet in the next future.

1.1 Exploration aims

The high number of mission explorations to Mars clearly points out the importance of Mars within the Solar System. Thus, the question is: "Why this great interest in Mars exploration?"

The interest in Mars is due to several practical, scientific, and strategic reasons. In the practical sense, Mars is the most accessible planet in the Solar System [1]. It is the second closest planet to Earth, besides Venus, averaging about 360 million kilometers apart between the furthest and closest points in its orbit. Earth and Mars feature great similarities. For instance, both planets rotate on an axis with quite the same rotation velocity and tilt angle. The length of a day on Earth is 24 h, while slightly longer on Mars at 24 h and 37 min. The tilt of Earth axis is 23.5 deg, and Mars tilts slightly more at 25.2 deg [2]. Further, Earth and Mars revolve around the Sun with about the same revolution velocity. The former orbits at 30 km/s and the latter at 24 km/s. A year is 365 days on Earth and almost double that at 687 Earth Days on Mars [2]. Both Earth and Mars have four seasons each. Severe dust storms occur during the summer in the Mars' southern hemisphere. They are so strong that block most of the surface from view by satellites. During the fall, in the Mars' polar regions, crystals of carbon dioxide (CO_2) form and so much of the atmosphere gets absorbed that atmospheric pressure drops up to 30% as seasons transition from fall to winter [2].

From the scientific point of view, it is worth noting that exploring Mars provides the opportunity to possibly answer origin and evolution of life questions and could someday be a destination for survival of humankind. In fact, the red planet is a stony body with atmosphere, like Earth, with the same age, yet with only half the diameter of Earth, and with similar geological structures, as cold and desert-like

IntechOpen

surfaces, mountains and volcanoes, lava plains, cratered areas, giant canyons, wind-formed features, ancient riverbeds and, a very important thing, the presence of water. Therefore, Mars could provide several information about what the Earth beginning was and, especially, what the future expected to Earth will be, if climate changes arise and persist.

Finally, in the strategic sense, exploring Mars demonstrates political and economic leaderships of a nation, improves the quality of life on Earth, inspires young generations, and helps learn about our home planet.

The first successful approach to the red planet was with the flyby of the US Mariner 4 spacecraft (see **Figure 1**), on July 15, 1965 [1].

The flyby mission allowed getting the Mariner 4 space probe very close to Mars so that it was able to collect photos of the red planet in passing, see **Figure 2**. They were first Mars images ever returned to Earth [3].

Since then, more than 50 missions were planned and partially successfully accomplished. The Mars exploration, in fact, is characterized by a high failure amount, especially the early attempts. Roughly 30% of all Mars missions failed before completing their goals and some failed even before their observations could start. As is well known, Mars missions represent a great success for aerospace engineering because of they feature several complex phases, like launch, interplanetary journey, approach, entry and descent flights, and landing; each one is unique and demands great efforts. These exploration missions answered several fundamental questions, such as dispelling the myth of canals that evidenced an ancient civilization and investigated the antique riverbeds present on the surface, which suggests the presence of liquid water and maybe of past life forms, which may lie hidden below the planet's forbidding exterior.

1.2 Mars missions timeline

The timeline of the most important Mars exploration missions is herein reported [5].

1965—Mariner 4 (see **Figure 1**) passes very close to Mars (within nearby 10,000 km) and provides the first closest images of the surface [5].

1969—Mariners 6 and 7 pass at around 4000 km from the planet and send several data of Mars atmosphere and surface [5].

November 3, 1971—Mariner 9 orbits on November 24. It is the first US probe orbiting around a celestial body other than Earth. Mariner 9 returns detailed data on the planet's atmosphere, maps its surface, reveals Martian topography, and captures many more images of the red planet [5].

December 2, 1971—The USSR's Mars 3 orbiter (see **Figure 3**) returns some 8 months of data that reveal much about the planet's topography, atmosphere, weather, and geology [5].

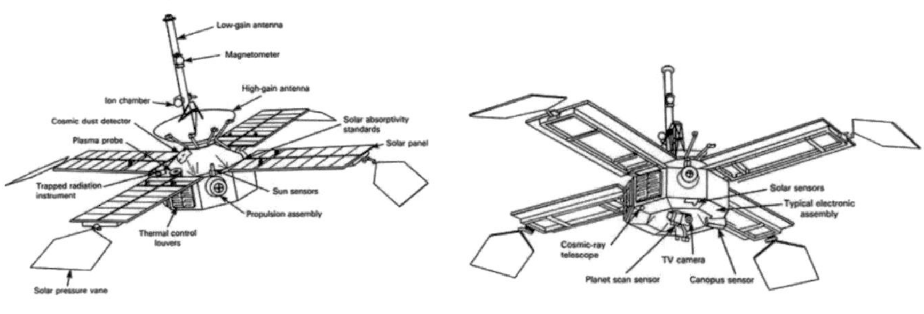

Figure 1.
The Mariner spacecraft. Credit: NASA [1].

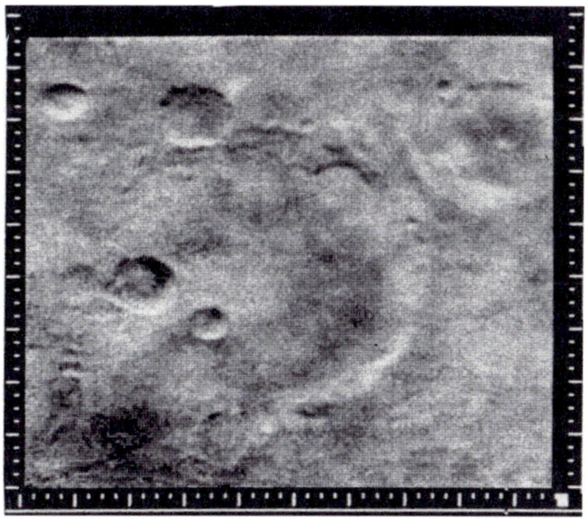

Figure 2.
Photo of Mariner crater on Mars taken by Mariner 4. Credit: NASA/JPL [4].

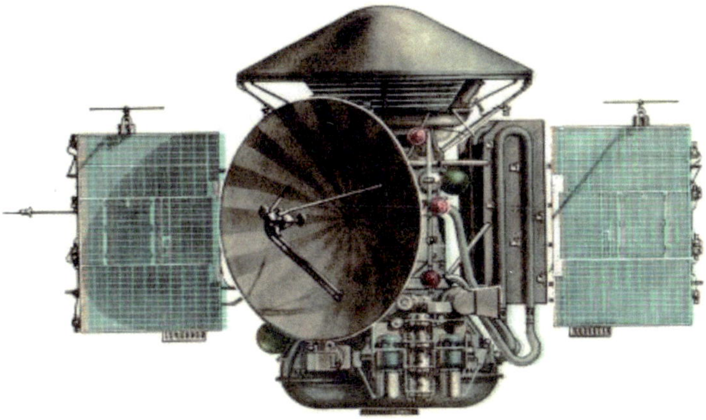

Figure 3.
The Mars 3 spacecraft. Credit: NASA [6].

But the major success is that for the first time, the Mars 3 lander successfully touches the planet's surface, see **Figure 4**. But, the lander sends data for only 20 s [5].

July and August 1973—The USSR launches Mars 4, 5, 6, and 7, but only Mars 6 lands [5].

July 19, 1976—Viking 1 (see **Figure 5**) arrives at Mars. It was launched on August 20, 1975. Viking 1's lander (see **Figure 6**) reaches the Mars surface on July 20 [5]. The primary mission objectives were to obtain high resolution images of the Martian surface, characterize the structure and composition of the atmosphere and surface, and search for evidence of life [8].

September 3, 1976—The lander of Viking-2 reaches the surface of the red planet. Viking 1 and 2 spacecrafts continues to send back data as late as 1982 [5]. They capture extraordinary images of the Mars surface that astound the public and excite scientists. The landers also conduct biological experiments on soil to search for evidence of life in space—but their results are inconclusive, though tantalizing [5].

September 25, 1992—NASA launches the Mars Observer (MO) [5].

September 1997—The Mars Global Surveyor (MGS) reaches Mars and begins its orbit up to the end of mission on November 2, 2006 [5]. NASA's MGS mission

Figure 4.
The Mars 3 lander. Credit: NASA [6, 7].

Figure 5.
The Viking spacecraft. Credit: NASA [8].

Figure 6.
The Viking lander. Credit: NASA [8].

objective is mapping and studying the Mars' surface. The mission provided insights into the dynamic changing seasons and shifting weather, including its strong dust storms [5].

July 4, 1997—The Mars Pathfinder lands on Mars and the rover Sojourner starts exploring the planet and sends back images to Earth for 4 months, see **Figure 7**. Sojourner proves that it is possible to land a free-moving rover vehicle that travels around the Mars surface, thus collecting scientific data, such as images, weather observations, and chemical soil analyses [5].

December 11, 1998—NASA launches the Mars Climate Orbiter (MCO) but unfortunately is lost on arrival in September 1999 [5].

January 3, 1999—NASA launches the Mars Polar Lander (MPL) but the spacecraft is not able to communicate with the ground control on December 3 [5].

October 24, 2001—The Mars Odyssey Orbiter (MOO) reaches Mars [5].

June 2, 2003—The Mars Express Orbiter (MEO) is launched by the European Space Agency (ESA), with the Beagle 2 lander [5]. MEO spacecraft successfully orbits around Mars, where it studies the planet for 2 years. Unfortunately, the Beagle 2, planned to land on December 25, 2003, never makes contact. January 16, 2015—NASA declares that the Beagle 2, has been found on Mars with solar panels not fully open upon landing. This probably not allowed communication with ground control [5].

June 10, 2003—The rover Spirit is launched [5].

July 7, 2003—NASA launches the rover Opportunity. Both Spirit and Opportunity (see **Figure 8**) belong to the NASA's Mars Exploration Rover Mission [5].

January 3, 2004—Spirit lands on the red planet and begins transmitting images of the Mars surface [5].

January 15, 2004—The rover Spirit leaves its lander and starts to study the rocks and soil for evidences of water [5].

January 25, 2004—The rover Opportunity reaches the Mars surface and starts its exploration mission that lasts 33 months longer than originally designed [5].

Figure 7.
Pathfinder lander camera image of Sojourner atop the "Mermaid Dune" on Sol 30. Credit: NASA [9].

Figure 8.
An artist's concept portraying a NASA Mars exploration rovers opportunity and Spirit on the Mars surface. Credit: NASA [10].

February 2, 2004—Both Spirit and Opportunity rovers are fully operational and explore Mars surface by gathering soil samples and images, to find sign of the past Martian environment [5]. The last communication is heard from Spirit on March 22, 2010, while Opportunity is still operational. June 10, 2018, the last communication is heard from Opportunity in Perseverance Valley. Finally, on February 13, 2019, NASA announces the Opportunity mission complete. Spirit and Opportunity rovers cover several kilometers and return a lot of high-resolution photos. Probing soil and rocks, they also use their fully equipped lab modules to conduct geology tests on the surface and below it [5].

March 10, 2006—The Mars Reconnaissance Orbiter (MRO) reaches the red planet and starts to scan Mars for more signs of water. MRO features the most powerful camera ever to leave Earth, called HiRise [5].

August 4, 2007—NASA launches the Mars Phoenix Lander (MPL). It lands near Mars northern pole on May 25, 2008. [5]. Phoenix belongs to the Scout program for low-cost and small spacecraft of NASA and aims to analyze soil samples and scan the atmosphere. In November 2008, the MPL stops communications after the mission competition which last for 149 days. The mission is expected to last for 90 Martian solar days [5].

Finally, on September 10, 2010, scientists release data from the MPL where it appear that water has been weathering planet surface during history. Data also points out that CO_2 atmosphere has been supplied by geologically recent volcanic eruptions, suggesting the presence of a possible ongoing activity [5].

April 6, 2011—NASA shows Curiosity, see **Figure 9**. It is a rover bigger than Spirit and Opportunity, with about 900 kilograms weight and the dimensions of a small utility car [5]. Curiosity is launched on November 26, 2011 and successfully lands on Mars on August 6, 2012. June 7, 2018—NASA declares that an organic matter has been detected by the rover in the Gale crater soil samples. Curiosity has also found methane in the atmosphere [5].

September 21, 2014—NASA's Mars Atmosphere and Volatile Evolution (MAVEN) mission space probe orbits around the red planet with the aim to study the higher atmosphere in order to understand the Mars' past climate changes and habitability [5].

September 24, 2014—India's Mars Orbiter Mission (MOM) orbits around Mars. India is the first Asian nation to reach Mars and the first country to successfully accomplish the mission on its first attempt [5]. The spacecraft continues to operate after its 160-day mission.

Figure 9.
NASA Mars exploration rovers Curiosity. Credit: NASA [11].

March 14, 2016—ESA launches Trace Gas Orbiter (TGO) with a lander called Schiaparelli, within the program Exobiology on Mars (ExoMars) [5]. TGO successful enters Mars' orbit, while Schiaparelli fails a soft landing and crashes. TGO detects atmospheric gases to prove the presence of life on Mars.

May 5, 2018—NASA launches Interior Exploration using Seismic Investigations, Geodesy and Heat Transport (InSight) [5]. It is the first robotic explorer designed to study the interior of Mars. InSight lands just north of the planet's equator on November 26, 2018, joining five other NASA space probe working on and above Mars.

July 25, 2018—The results of a study of Italian Space Agency (ASI) scientists indicates the presence of a liquid water lake below the southern polar ice cap of Mars. This results have been achieved using the Mars Advanced Radar for Subsurface and Ionosphere Sounding (MARSIS) instrument of ESA [5].

1.3 Next planned robotic missions

The end of July 2020 will see even three robotic exploration missions. Il will be multiple launches never seen before in the history of space exploration.

July 20, 2020—The United Arabs Emirates Mars Mission is a planned space exploration mission which aims at entering in Mars' orbit, the probe is called Hope.

July 23, 2020—Tianwen-1 is a planned mission by China. The space probe features an orbiter and a lander with a rover. Mission aims are to find out sign of both current and past life, and to investigate the Mars environment.

July 30, 2020—The NASA Mars 2020 mission belongs to Mars Exploration Program. It comprises the Perseverance rover to investigate Mars' surface geological processes and history, including the appraisal of its past habitability, ancient life forms, and the potential for the preservation of biosignatures within accessible geological materials.

1.4 The next frontier: human-rated missions

So far, the remotely unmanned exploration missions provided a huge increase in knowledge about the Martian system, especially for what concerns the understanding of its geology and habitability potential. But the red planet still represents the next frontier of space exploration.

The new goal, in fact, is to send astronauts to Mars with acceptable cost, risk, and performance. Proposals for human-rated missions have been made throughout the space exploration history and have also been an inspiration since the earliest days of Space Science. For instance, for Robert H. Goddard, the idea of reaching the red planet was fundamental to overcome all the difficulties he met in his research studies on the Space Flight.

Human exploration was identified as a long-term goal in the US space exploration vision. Conceptual works and plans of NASA focus on missions typically being stated as taking place after about 10–30 years from the time they are drafted. On September 28, 2007, NASA's then-administrator Michael D. Griffin stated that the US aims to send men on Mars by 2037. The Orion crew exploration vehicle (CEV) would be used to send a human mission to the Moon by 2024 within the Artemis program as a steppingstone to a Mars expedition [12]. A flight to Mars could follow this program.

Currently, active plans foresee the landing on Mars of a crew made of two up to eight astronauts that would visit the red planet for a period of a few weeks or a year. The aim is exploration at a minimum, with the possibility of sending settlers and terraforming the planet for the permanent colonization of Mars or exploring its moons Phobos and Deimos as well. On October 8, 2015, NASA unveiled its official plan for human missions to Mars, called Journey to Mars, leading up to fully sustained colonization. The plan develops through three phases, namely Earth Reliant phase, Proving Ground phase, and Earth Independent phase. During the first phase (Earth Reliant), the International Space Station will be used to validate deep space technologies and study the effects of long duration space missions on the human physiology. The second phase (Proving Ground) will focus on the research/scientific activities away from Earth and to be accomplished into cislunar space. Activities in this phase are to test deep space habitation facilities, and validate capabilities needed for human exploration of Mars. Finally, the third phase (Earth Independent) features the transition to independence from Earth resources. This last stage comprises long-term missions on the lunar surface, which leverage surface habitats that only require routine maintenance, and the harvesting of Martian resources for fuel, water, and building materials. Anyway, it is expected that Earth's independence could take several decades.

Crewed mission to Mars is extremely challenging both from the engineering and human point of views. A safe, reliable, and affordable mission design must include the shortest interplanetary journey, a safe transportation system with radiation shielding, centrifugal artificial gravity, in-transit consumable resupply, and an orbiter and a lander which can return.

As far as human factor is concerned, it is worth noting that a not negligible mission aspect is the psychological effort. A typical mission would last at least 18 months. Therefore, for a crew staying for this long time and very far from Earth (for the first time in the history) continuously subject to a hostile environment could be extremely challenging from a psychological point of view. Several experiments are planned or yet executed by Space Agencies in order to make sure that astronauts are prepared mentally and physically for the demands of long exploration missions. For example, ESA undertook a cooperative project with the Russian Institute for Biomedical Problems (IBMP) in Moscow, called Mars500. This test began in February 2010 and the end of the isolation period was November 4, 2011 [13]. Further, on August 28, 2015, NASA funds a year-long experiment, with six scientists, to investigate the effects of long Mars mission. The crew lived in a bio dome on Mauna Loa mountain, Hawaii, with partial connection to the outside world and were only allowed outside if they were wearing spacesuits [14].

For the time being, Mars still inspires next generations of scientists (physicists, engineers, mathematicians, and so on). It seems still far from human's exploration, but great efforts are made, ongoing and planned for the next future.

A day will arrive when a new small step will be moved forward for a giant leap for mankind.

Author details

Giuseppe Pezzella* and Antonio Viviani
Department of Engineering, University of Campania "L. Vanvitelli," Aversa, Italy

*Address all correspondence to: giuseppe.pezzella@unicampania.it

References

[1] https://mars.nasa.gov/#mars_exploration_program/3

[2] https://mars.nasa.gov/all-about-mars/facts/

[3] https://www.hq.nasa.gov/office/pao/History/SP-4212/p164.html

[4] http://www.jpl.nasa.gov/history/60s/Mariner4_1965.htm

[5] https://edition.cnn.com/2013/10/21/world/mars-exploration-fast-facts/index.html

[6] http://nssdc.gsfc.nasa.gov/image/spacecraft/mars3_iki.jpg

[7] http://nssdc.gsfc.nasa.gov/image/spacecraft/mars3_lander_vsm.jpg

[8] https://nssdc.gsfc.nasa.gov/planetary/viking.html

[9] http://photojournal.jpl.nasa.gov/index.html

[10] http://photojournal.jpl.nasa.gov/catalog/PIA04413

[11] http://photojournal.jpl.nasa.gov/catalog/PIA19920

[12] https://www.nasa.gov/specials/artemis/

[13] https://www.esa.int/Science_Exploration/Human_and_Robotic_Exploration/Mars500/Mars500_study_overview

[14] https://earthobservatory.nasa.gov/images/92630/living-the-mars-life-on-mauna-loa

Chapter 2

Autonomous Navigation for Mars Exploration

Haoting Liu

Abstract

The autonomous navigation technology uses the multiple sensors to percept and estimate the spatial locations of the aerospace prober or the Mars rover and to guide their motions in the orbit or the Mars surface. In this chapter, the autonomous navigation methods for the Mars exploration are reviewed. First, the current development status of the autonomous navigation technology is summarized. The popular autonomous navigation methods, such as the inertial navigation, the celestial navigation, the visual navigation, and the integrated navigation, are introduced. Second, the application of the autonomous navigation technology for the Mars exploration is presented. The corresponding issues in the Entry Descent and Landing (EDL) phase and the Mars surface roving phase are mainly discussed. Third, some challenges and development trends of the autonomous navigation technology are also addressed.

Keywords: autonomous navigation, deep space exploration, inertial navigation, celestial navigation, visual navigation, integrated navigation

1. Introduction

The Mars is an attractive planet in the space, which looks red, faraway, and mysterious. It is believed that the Mars is the second planet, which is suitable for people to live [1]. After many years of probing, it has been found that the Mars has plenty of minerals, its average surface temperature is about −60°C, it has thin atmosphere, and it even has water [2]. The scientific fiction is if we deploy some solar reflectors in Mars' orbit and build many greenhouse gas factories in its surface, the temperature there will increase gradually, and then the thick atmosphere [3] will be created, the ice in its polar and underground will melt, and finally, the primitive life may appear. Nowadays, to land that red planet and build new base, there is the competitive focus of aerospace technology for many countries. However, it is known that the distance between the Earth and the Mars is between 5.5×10^7 and 4×10^8 km, and it will cost us almost 7 months to reach that place. To reach that red planet in the vast and boundless universe safely, all the solutions point to the navigation technology [4].

In general, the flight mission of Mars exploration [5] can be divided into four phases. The first one is the launch phase. The rocket will send the Mars prober into the space and escape from the gravitational constraint of the Earth. The second one is the cruise flight phase. The prober will fly to the Mars by several times of orbit transfer [6]. The third one is the Entry, Descent, and Landing (EDL) phase [7]. This is the most dangerous flight stage in this mission because of the influences of the thin Mars atmosphere, the Mars gravitation, and its complex terrain. The last one is the roving phase. After a safe landing, the rover will walk on the surface of

Mars and look around here and there. Clearly, different navigation technologies should be employed in different flight phases. The satellite navigation [8], the inertial navigation [9], the celestial navigation [10], the visual navigation [11], and the integrated navigation [12] are the most popular navigation techniques in the aerospace engineering research field. **Figure 1** presents the sketch map and an example of Mars exploration. In **Figure 1**, (a) is the sketch map of Mars exploration flight mission, and (b) gives out a kind of landing method in the EDL phase.

Many research works have been done in the aerospace navigation research field. In [13], the authors proposed a new method, which could use the impulses in the solar light and the time difference of arrival method to improve the navigation accuracy. In [14], an inertial measurement unit and the orbits/aerostats beacons-based integrated navigation scheme were developed. The range and the Doppler information were integrated in the extended Kalman filter. In [15], a navigation method, which could use the inter-satellite link and the starlight angle, was simulated. The prober could determine its position by communicating with other space crafts and observing the stars. In [16], a norm-constrained unscented Kalman filter-based navigation scheme

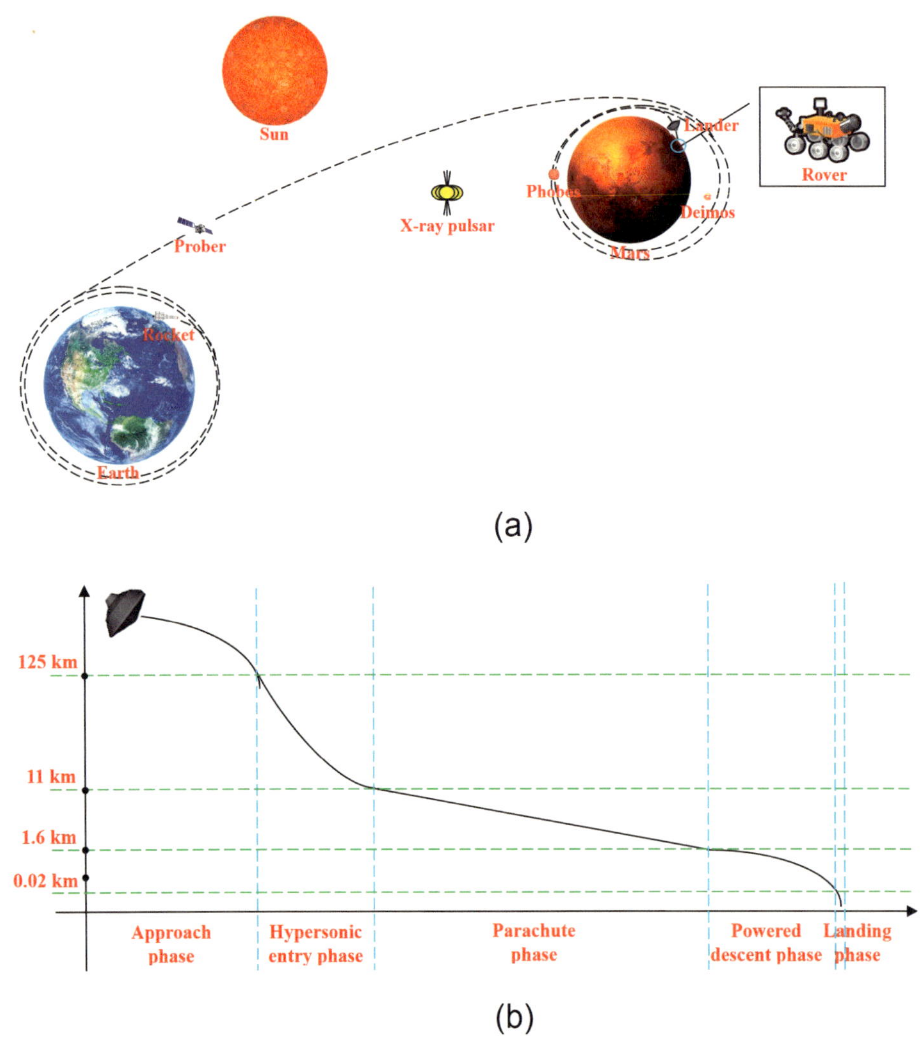

Figure 1.
The sketch map and an example of Mars exploration: (a) is the sketch map of the flight mission; (b) is a landing example in the EDL phase.

was proposed. This method could be used for the spacecraft attitude quaternion estimation during the Mars powered descent. After a comparable study of the popular navigation methods, it can be found that the navigation technique has the development trend of high accuracy, multi-function, mini-size, low power consumption, and long service life. In addition, the Artificial Intelligence (AI) technique [17] also has the potential application value in the navigation research field.

2. The key autonomous navigation techniques

Different from the traditional navigation techniques, the autonomous navigation [18] can realize the perception, the computation, and the decision making of the flight trajectory all by the aerospace prober itself. No exterior information is needed. Sometimes the autonomous navigation system can also emit some state information to the ground for the safety monitoring purpose. Many methods can be used to assist the autonomous navigation flight. The popular methods include the inertial navigation [19], the celestial navigation [20], the visual navigation, and the integrated navigation. **Figure 2** shows the sketch map of the popular autonomous navigation methods.

2.1 The inertial navigation

The inertial navigation can measure the acceleration of prober and use the integral computations to estimate its transient flight speed and spatial position. During that process, no energy will be radiated from the inertial navigation device, and no exterior auxiliary information is needed. The inertial navigation adopts the devices of the accelerometer [21] and the gyroscope [22] to realize the state estimation. The accelerometer [23] is a device, which can measure the acceleration of the carrier. In general, it can be classified into the linear accelerometer and the angular accelerometer. In the practical application, the linear accelerometer is commonly used because of its high precision and excellent stability. Many accelerometers have been designed, such as the piezoelectric accelerometer, the piezoresistive accelerometer, or the capacitive accelerometer. **Table 1** shows a kind of classification of the linear accelerometer. The gyroscope is a device, which can measure the rotation rate. **Table 2** illustrates some gyroscope products. They are the mechanical rotor gyroscope, the electrostatic suspended gyroscope, the vibratory gyroscope [24], the laser gyroscope, and the fiber optic gyroscope [25].

The inertial navigation is the most mature and popular method in the navigation research field. Comparing with other navigation techniques, the inertial navigation at least has three advantages. First, it is independent to the environment.

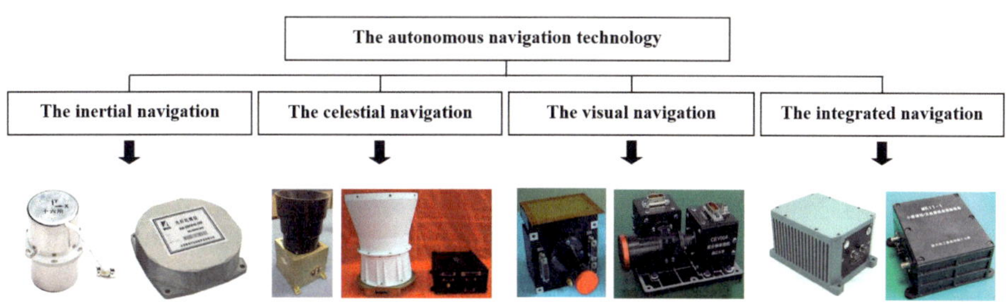

Figure 2.
The popular autonomous navigation methods.

No.	The name	The basic principle	The application
1	The piezoelectric accelerometer	The principle of piezoelectric effect	The quartz crystal or the piezoelectric ceramics accelerometer
2	The vibrating string accelerometer	The relationship between the resonant frequency of string and the corresponding acceleration	The electromagnetic oscillator-based accelerometer with the string
3	The vibrating beam accelerometer	The relationship between the resonant frequency of beam and the corresponding acceleration	The electromagnetic oscillator-based accelerometer with the beam
4	The optical accelerometer	The principle of optical interference effect	The grating accelerometer, the fiber-optic accelerometer, or the polymer accelerometer
5	The pendulous accelerometer	The relationship between the rotation angle and the corresponding acceleration	The liquid-floated pendulous accelerometer, the flexure hinged pendulous accelerometer, and the pendulous integrating accelerometer

Table 1.
A kind of classification of the linear accelerometer.

No.	The name	The basic principle	The application
1	The mechanical rotor gyroscope	The Newton's mechanics laws: the mechanical rotor can maintain its rotation orientation regardless of its base movement	The gas-floated mechanical rotor gyroscope, the liquid- floated mechanical rotor gyroscope, and the ordinary mechanical rotor gyroscope
2	The electrostatic suspended gyroscope	The ball rotor, which works in a vacuum electric field, can maintain its rotation orientation regardless of its shell movement	The electrostatic suspended gyroscope with a solid rotor, the electrostatic suspended gyroscope with a hollow rotor, and so on
3	The vibratory gyroscope	The Coriolis effect, which is observed from a vibrating structure	The tuning fork gyroscope, the hemispherical resonator gyroscope, and so on
4	The laser gyroscope	The Sagnac effect	The four-mode differential laser gyroscope, the mechanically dithered ring laser gyroscope, and so on
5	The fiber optic gyroscope	The Sagnac effect	The interferometric fiber optic gyroscope, the resonator fiber optic gyroscope, and the Brillouin fiber optic gyroscope

Table 2.
A kind of classification of the gyroscope.

It can realize the all-weather and the all-time working modes in any places. Second, it can provide the position, the speed, the course, and the attitude information of carrier accurately. And its data updating ratio is also high. Third, it has the good performances in the anti-interference and the low system noise. Clearly, the inertial navigation also has some shortcomings. For example, according to its navigation principle, its navigation error will be accumulated. Another problem is that the inertial navigation also cannot give out the time information. Recently, the Micro Electro Mechanical Systems (MEMS) technology [26] has met its great development opportunity. The MEMS technology can realize the optoelectronics product manufacturing in a microscopic degree. It has been used to improve the product performance of the vibratory gyroscope, the laser gyroscope, and the fiber optic gyroscope in recent years. In future, the miniaturization, the high precision, and the high reliability processes will be developed in the inertial navigation system.

2.2 The celestial navigation

The celestial navigation uses the space object, such as the sun, the moon, or the other stellar [27] as a reference to guide the flight direction and the flight attitude of the prober. Clearly, the star sensor [28] should be employed in this method. As a kind of optical measurement device, besides the visible light camera, the infrared camera, the ultraviolet camera, the X-ray camera, the γ-ray camera, and so on can all be used. Many methods have been designed for the celestial navigation, that is, the angle measurement-based navigation [29], the distance measurement-based navigation [30], and the speed measurement-based navigation [31]. The angle measurement-based navigation uses the angles among the sun, the other planets, the planet satellite, the asteroid, or the comet to carry out the autonomous navigation. The distance measurement-based navigation mainly employs the detection result of the arrival time difference between the X-ray pulsar and the standard pulse of the solar system center to estimate the navigation information and the time offset. In addition, the speed measurement-based navigation utilizes the optical Doppler effect [32] to implement the navigation. **Table 3** presents the popular methods of the celestial navigation.

The celestial navigation at least has three advantages. First, as a kind of optic measurement device, most of the celestial navigation systems are immune to the traditional electromagnetic interference problem [33]. Also, their working reliabilities are comparable high. Second, its navigation accuracy is good. Because the spatial distance between the celestial body and the prober is really large, the navigation precision can be very high if the star sensor performance and the corresponding data processing algorithm are proper. Third, most of the celestial navigation devices can work all-time long as long as the corresponding celestial bodies are visible. The celestial navigation also has some shortcomings. The biggest problem is that some devices cannot be used in the complex light environment. For example, the strong environment light or the cosmic rays will affect the application effect of these optic devices seriously. Another drawback is that some celestial navigation devices cannot give out the sequential output because of the target star occlusion problem. Comparing with the inertial navigation technique, the celestial navigation is still immature, and most techniques do not get the test of the practical flight mission. In future, the new modeling technique of the orbit function [34], the new measurement theory of the star, and the new estimation method of the navigation information can all be researched extensively to improve the application effect of the celestial navigation.

2.3 The visual navigation

The visual navigation uses the imaging sensors to accomplish the navigation task. In this chapter, the visual navigation means to use the visual sensors and the corresponding algorithms to the autonomous robot vehicle or the unmanned aerial vehicle for the navigation purpose. In many cases, the visual navigation is used as the direction guidance or the obstacle avoidance device. Without loss of generality, the imaging sensors in the visual navigation include the visible light camera, the infrared or the near-infrared camera, the multispectral camera, the laser imaging camera [35], and even the ghost imaging (i.e., the quantum imaging) camera [36].

The imaging system can be realized by the monocular vision system [37], the binocular vision system, the multi-sensor vision system, and the depth vision system [38]. In engineering, the narrow field camera, the wide field camera, and the panoramic camera can always be used. **Table 4** presents the popular visual navigation techniques. Different from the traditional navigation methods, the visual navigation can capture 2D image data; thus, the image processing and the pattern recognition techniques can be used for its navigation information analyses and estimations. For example, the region-based matching, the feature-based matching (point, line, or region feature), and the semantics-based matching can all be developed [39].

Clearly, the visual navigation has lots of application advantages. First, the imaging data have the abundant details, they are easy to be understood by people, and their processing results are also intuitive. Second, the imaging sensor can be used under a flexible working mode, such as the active mode or the passive mode. The active mode means the imaging sensor system will project some light to the observed target [40] to improve its visibility. The projected light can be the visible light, the infrared light, or even the laser in certain spectral bands. The passive mode indicates that the imaging sensor system will not project any light to the exterior environment. Third, the navigation accuracy is high. In many cases, the image-based navigation can realize the high precision navigation in centimeter degree or even more. This excellent performance can guarantee it to be used in some complex

No.	The name	The basic principle	The application
1	The sextant	The angle between any two typical objects can be used as a navigation reference	The celestial sextant
2	The celestial sensor	The spatial positions of some planets can be used as a navigation reference	The star sensors, such as the X-ray pulsar sensor, the asteroid sensor, the Mars sensor, the sun sensor, and the planetary photographic instrument
3	The horizon sensor	The horizons of some planets can be used as a navigation reference	The Earth horizon detection device, the Mars horizon detection device, and so on

Table 3.
A kind of classification of the celestial navigation technique.

No.	The name	The basic principle	The application
1	The map-based navigation	The digital map construction and the feature matching	The navigation method in a known environment, it can be implemented by the absolute positioning and the relative positioning techniques, and so on
2	The optical flow-based navigation	The feature detection and the motion information estimation	The translation motion estimation method and the rotation motion estimation method
3	The appearance-based navigation	The image feature matching	The terrain matching and the landmark matching
4	The integrated visual navigation	The information fusion of different visual navigation techniques	The visual light camera and the infrared camera-based navigation, the visual light camera and the laser radar-based navigation, and the visual light camera and the depth camera-based navigation

Table 4.
A kind of classification of the visual navigation technique.

terrain environments. Obviously, the visual navigation also has some drawbacks. For example, its data processing amount is very large. Comparing with the data processing of the inertial navigation, the computation of the image processing algorithm is really complex; as a result, the visual navigation needs its processing unit to contain the high data processing performance and the large power consumption. Another problem is that the visual sensor is sensitive to the environment light [41]. All kinds of atmosphere phenomena will create great negative influences on its processing result.

2.4 The integrated navigation

The integrated navigation uses the combination of two or more than two basic navigation methods above to realize the high precision direction guidance and the attitude estimation of carrier. Its key techniques include the computer-based feature extraction, the information fusion, and the time synchronization [42]. The integrated navigation can make up the disadvantage of the sole navigation method, realize the seamless navigation, and achieve the information redundancy designing of the navigation system [43]. Its system reliability and the navigation accuracy can be improved a lot. Currently, the most successful integrated navigation is the inertial plus the Global Positioning System (GPS)-based navigation method, that is, the combination of the inertial navigation and the satellite navigation. Regarding the autonomous navigation technique for the deep space exploration, **Table 5** presents the popular integrated navigation methods. From **Table 5**, any two basic navigation methods can be used for the integrated navigation application. The integrated navigation also has some drawbacks. For example, its complex system design, its big size, and its high power consumption are all the problems for the spacecraft application.

2.5 Some other autonomous navigation methods

Some other autonomous navigation methods can also be used for the Mars exploration. First, the radio-based navigation [44] is a choice. Traditionally, the radio-based navigation uses the satellites or the beacons on the ground to provide the prober some guidance information. Clearly, this method is not an autonomous technique. However, if we look both the satellite and the prober as a whole that means the Mars exploration prober system includes both the lander and the navigation satellites, the radio-based navigation can be considered. Second, the bionic navigation method [45] is another choice. The bionic navigation imitates the animals' sensory processing to accomplish the navigation. The stereoscopic vision, the auditory

No.	The integrated navigation method	The application
1	The inertial navigation and the celestial navigation	Three accelerometer, three gyroscope, and the star sensors (the observation from orbit)
2	The inertial navigation and the visual navigation	Three accelerometer, three gyroscope, the visible light camera, and the laser radar
3	The visual navigation and the celestial navigation	The visible light camera, the laser radar, and the star recognition sensor (the observation from ground)
4	The inertial navigation, the celestial navigation, and the visual navigation	Three accelerometers, three gyroscopes, the star sensors, and the celestial body feature recognition sensor

Table 5.
A kind of classification of the integrated navigation technique.

sensation, the olfactory sensation, and the tactile sensation can all be utilized in this method. Lost of new sensors can be developed for this navigation technique. Third, the Mars atmosphere model-based navigation [46] can also be considered. For example, a kind of Mars flush air data system has been used to assist the estimations of the dynamic pressure, the Mach number, and the impact angle of Mars prober. By using all the new navigation theories and methods, the final navigation accuracy can be improved definitely.

3. The autonomous navigation for Mars exploration

3.1 The autonomous and the nonautonomous navigations in Mars exploration

As we have stated, the Mars exploration mission can be divided into four phases, that is, the launch phase, the cruise flight phase, the EDL phase, and the roving phase. **Table 6** illustrates the comparisons between the autonomous navigation method and the nonautonomous navigation technique during four phases. In fact, both the autonomous navigation and the nonautonomous navigation methods are needed in the deep space exploration. In **Table 6**, regarding the nonautonomous navigation methods, the GPS, the Beidou satellite system, the GLONASS, and the Galileo system are the most famous satellite navigation systems in the world. The radio-based method can carry out the navigation by the means of the Deep Space Network (DSN)-based system [47] or its airborne radar equipment. The DSN is a land-based tracking, telemetry, and control network for the satellite navigation, control, and observation in the deep space. The giant radio emission equipment, the X-ray telescope, and some communication terminals are deployed in different sites on Earth or in orbit to assist the deep space navigation application.

The navigation in the EDL phase faces lots of challenges. First, the complex atmosphere and weather in Mars will create uncertainty to the navigation task. The attenuated atmosphere will cause the deficiency of the aerodynamic force [48]; no accurate environment model and aerodynamic model can be used; the gust and the sandstorm will always affect the land precision; and the descent speed is fast, and the blackout zone [49] still exists, which can affect the normal communication and so on. Second, the long-time space flight will rise changes and damages to the mechanical structure and the sensors [50] of prober. The cosmic ray, the extreme temperature, and the space debris will influence the health state of prober. Third, the time delay, which is larger than 10 minutes from Earth to Mars, also indicates that people cannot monitor its flight state in real time. To conquer these problems to some extent, the autonomous navigation can be used during the EDL phase. In the entry stage, both the inertial navigation and the celestial navigation can be used; in the descent stage, the inertial navigation can be considered; and in the landing stage, the inertial navigation and the visual navigation can all be employed.

The visual navigation in the roving phase also has some problems. First, since lots of optic devices are utilized, the complex environment light and the dust in air will influence their working states [51] seriously. Because the Mars surface has no liquid water and no vegetation, the wind with the sand can be commonly met. The lens of imaging sensor may be contaminated. Second, the low surface temperature in Mars also is a problem. Most of storage batteries cannot work under a normal state in that low temperature. The strength of materials will also be changed. Third, the poor performance of the imaging sensor can determine the limited computational effect of the image processing algorithm. To control the weight of launch load, the high performance camera with big size cannot be sent to Mars. Thus, how to use the images come from the low resolution imaging sensor for the data analysis

No.	The flight phase	The autonomous navigation methods	The nonautonomous navigation methods
1	The launch phase	The inertial navigation and the celestial navigation	The satellite navigation and the radio navigation
2	The cruise flight phase	The inertial navigation and the celestial navigation	The radio navigation, and so on
3	The EDL phase	The inertial navigation, the celestial navigation, and the visual navigation	The radio navigation, and so on
4	The roving phase	The inertial navigation, the celestial navigation, and the visual navigation	The radio navigation, and so on

Table 6.
The autonomous and the nonautonomous navigation methods in different flight phases of Mars exploration.

purpose is a challenge. To conquer these problems above, on one hand, the special materials and process will be used to improve the performance of the imaging sensor; on the other hand, some typical workflows and data fusion algorithm [52] should be designed according to the environment characters in Mars.

3.2 The autonomous navigation algorithms

Many navigation algorithms have been developed for the autonomous navigation application. Regarding the inertial navigation, lots of time series data can be collected from the accelerometer and the gyroscope. Because of the electrical radiation, the impact and the vibration, and the temperature drift, their measurement data are always contaminated by the system noise or the exterior noise. As a result, the denoising computation and the trending prediction are necessary. In general, the classic Kalman filter, the Extended Kalman Filter (EKF), the Unscented Kalman Filter (UKF) [53], and the particle filter [54] can be used to improve the data quality. The Kalman filter belongs to a kind of sequential filter technique, which needs no complex iterative computation. Its processing speed is fast, while its need of the data storage capacity is small. In other cases, the least square filter [55] can also be considered to solve the denoising issue. The least square filter is a kind of batch processing algorithm, which needs no priori information of the state variable; it has been used to solve the orbit estimation issue currently.

The celestial navigation uses the imaging sensor to observe the typical celestial body. Lots of data processing methods for 2D data can be employed. For example, the dim target detection can be used to identify the Phobos and the Deimos of Mars in space; the image denoising [56] can be considered for the noise inhibition; the spectral analysis can be employed to assist the atmosphere component analysis; and the image quality evaluation can also be used for the celestial body recognition. The visual navigation also adopts the imaging sensor to percept the environment. Thus, the related image processing algorithms [57], such as the camera calibration, the image enhancement, the image restoration, the image denoising, the feature detection, the feature matching, the image mosaic, and the image segmentation, can all be utilized. Recently, the Simultaneous Localization and Mapping (SLAM) technique [58] begins to be used for the map navigation in the narrow area. Its familiar methods include the Lidar-based SLAM and the visual-based SLAM. This technique can be used for the visual navigation in the Mars roving phase in future.

The integrated navigation can get the multivariate data; thus, the data fusion technique should be applied for its data processing. The data fusion can improve the data processing precision and reliability by the measurement of information reasoning and integration. In general, the data fusion algorithm includes the linear weighting-based method, the filtering with a Bayesian estimation-based method [59], the factor graph-based method [60], and the interactive multiple model fusion-based method. Among these methods, the linear weight-based method uses the sum of the weighted input data to carry out the fusion computation. The filtering with a Bayesian estimation-based method considers the prior information and the Bayesian statistical theory to implement the fusion calculation. The factor graph-based method employs the Bayesian network or the Markov model [61] to realize the data fusion. In addition, the interactive multiple model fusion-based technique uses two or more than two filters or the prediction model to accomplish data fusion. Clearly, the computation of data fusion can be realized by the parallel algorithm or the sequential algorithm.

3.3 The time service technique

The navigation system cannot leave the time service. The time service can provide a benchmark for the signal processing applications in the prober. Currently, the commonly used time service device in the satellite is the atomic clock [62]. In general, the atomic clock can include the cesium clock, the hydrogen clock, and the rubidium clock. Regarding the time service system, first, its time service precision should be high. For example, the time precision of some ground atomic clocks can reach 10^{-18} s. Second, the integrated performance of atomic clock [63] should be good. The service life span, the mechanical performance, the temperature sensitivity, and the working stability should be calibrated and tested strictly before it is sent to the space. Third, the time synchronization ability of atomic clock should also be excellent. Although the autonomous navigation will not use any exterior information for its navigation; however, it needs to transmit some important state information to the ground sometimes. As a result, the clock synchronization issue should be another key technique for the atomic clock.

3.4 The practical navigation application for Mars exploration

In a practical application, although the autonomous navigation has many merits; however, because of the high maturity and the good reliability, the nonautonomous navigation technique will be used in the Mars exploration mission as long as it is possible. Regarding the Mars exploration mission, the nonautonomous navigation method can include the satellite navigation [64], the ground radio navigation [65], and the Doppler radar navigation [66]. Here, the ground radio navigation is a typical application of the radio-based navigation. Lots of ground radio equipment or beacons can be used to provide the navigation information for the prober. The

No.	The flight phase	The integrated navigation method
1	The launch phase	The inertial navigation, the celestial navigation, the satellite navigation, the ground radio navigation, and so on
2	The cruise flight phase	The inertial navigation, the celestial navigation, the ground radio navigation, and so on
3	The EDL phase	The inertial navigation, the celestial navigation the visual navigation, the ground radio navigation, the Doppler radar navigation, and so on
4	The roving phase	The inertial navigation, the celestial navigation, the visual navigation, the ground radio navigation, the Doppler radar navigation, and so on

Table 7.
The application of the autonomous navigation and the nonautonomous navigation for the Mars exploration mission.

Doppler radar navigation also belongs to a typical application of the radio-based navigation. The Mars prober can use an active radar system to emit and receive signal to realize that navigation. Because of the extensive applications of these methods, both the ground radio navigation and the Doppler radar navigation are presented separately in this chapter. **Table 7** illustrates the integrated navigation method using both the autonomous navigation and the nonautonomous navigation for the Mars exploration. In fact, the nonautonomous navigation still plays a more important role in the flight mission than the autonomous navigation currently.

4. The challenges and the future works

4.1 The challenges

Although the explorations of the red planet have been performed for many years and lots of research plans have been proposed, many challenges still exist in the research field of autonomous navigation, including its accuracy, its reliability, and its service life of navigation system. **Table 8** shows some drawbacks of the autonomous navigation techniques. First, the accuracy issue can determine how precise when a navigation method is used for the Mars exploration task. Generally speaking, the accuracy of the integrated navigation is better than that of other sole navigation method. In the practical mission, both the autonomous navigation and the nonautonomous navigation should be used. Currently, the reported navigation accuracy of Mars landing mission is still not high. For example, the landing deviation of the curiosity rover (a Mars prober, which was launched by NASA in November 26, 2011) is about 10 km, while most of other landers could only reach the precisions from 100 to 300 km. The biggest uncertainty comes from the EDL phase. To improve the landing accuracy, the Mars atmosphere model, the orbit dynamics model, and the aerodynamics model should be researched elaborately in future.

Second, the reliability [67] is another problem. The reliability includes both the element reliability and the system reliability. The element reliability points to the probability of error free working state of each electronic or optic element. Also, the system reliability means the probability of error free working state of the whole navigation system. The redundancy designing degree of an aerospace system is also an evaluation index of the reliability. To improve the reliability, the environment experiments should be performed on ground to select the proper elements and test the robustness of the whole system. In general, the environment experiments include the temperature cycle experiment, the impact and the vibration experiment, the radiation experiment, the aging experiment, and the electromagnetic compatibility experiment. Third, the service life [68] of the Mars prober and its sensors also determine the result of the exploration mission. In many cases, the service life of a satellite can reach from several months to several years. The satellite healthy management is the new developed technique, which can extend the service life of satellite effectively. Its key techniques include the multiple sensors signal collection, the big data analyses, and the intelligent decision making and control.

4.2 The future works

Since the nature environment between the Earth and the Mars is similar, all the autonomous navigation methods developed in the Earth can be utilized in Mars. The first method should be the new-type inertial navigation techniques. The atom interferometric gyroscope [69], the nuclear magnetic resonance gyroscope [70],

No.	The autonomous navigation method	The shortcomings
1	The inertial navigation	The biggest problem is that the error can be accumulated. The navigation precision will be decreased with the time lapse.
2	The celestial navigation	The imaging sensor is easy damaged during the space flight. The practical application test of celestial navigation is still limited.
3	The visual navigation	The observation distance is small, and the environment light and the dust in air will affect the navigation accuracy seriously.
4	The integrated navigation	The system is complex and expensive, and its reliability is comparable low.

Table 8.
The drawbacks of the autonomous navigation techniques.

and the quantum gyroscope [71] can be developed and utilized for the Mars prober. The second method is the geomagnetic matching navigation method. In past, it is thought that the geomagnetic field in Mars is weak and disorder. Recently, some scholars have disclosed that parts of the local geomagnetic fields in Mars still could be used for navigation [72]. The third method is the gravity gradient navigation [73]. It is well known the gravitational acceleration has the diversity in different locations of the Earth. This is also true in Mars. If the gravitational acceleration map of Mars can be made in future, it will be used by the autonomous navigation technique in the red planet definitely. The fourth method is the bionic navigation [74]. Like an advanced animal, this navigation method can utilize the stereo vision, the auditory, and the tactile to realize the autonomous navigation.

The miniaturization is one of most important development directions of the autonomous navigation technique; thus, the MEMS should be emphasized. The MEMS is a technology, which can construct a system in the 1-to-100 μm degree. Sometimes its system size can have the outline in the millimeter degree. The MEMS has many merits for the aerospace application, such as the light weight, the low cost, the low power consumption, the long service life, the high reliability, the wide dynamic range, the fast response, and the easy installation. The micro-accelerometer and the silicon micromachined gyroscope are its representative products. In general, the inertial navigation system in a carrier has the application forms of the strap-down mode and the platform mode [75]. The former installs the inertial navigation sensors in the carrier directly, while the latter fixes the sensors in a platform and then puts the whole platform into the carrier. In many cases, the MEMS uses the strap-down mode to deploy its sensors. Clearly, the designing and the manufacturing of MEMS are not easy tasks. The environment test issue of MEMS is another problem. Because of its small size, the cosmic ray radiation, the low temperature, the low pressure, the zero gravity, and the impact and the vibration will also influence its service life and accuracy seriously.

Many works can be done in the navigation system and algorithm designing, and the corresponding navigation standards also need research works. **Figure 3** presents a kind of system designing for the navigation application in Mars. As we have stated, if we look both the navigation satellite and the prober as a whole, the satellite navigation can be looked as a kind of autonomous navigation method. With the assistance of the inertial and the celestial navigations, the navigation precision can be improved definitely. Some optimal algorithm can also be considered to improve the performance of the classic navigation method. For example, the genetic algorithm [76], the artificial bee colony algorithm [77], and the ant colony algorithm [78] can be used to improve the performance of the Kalman filter. Regarding

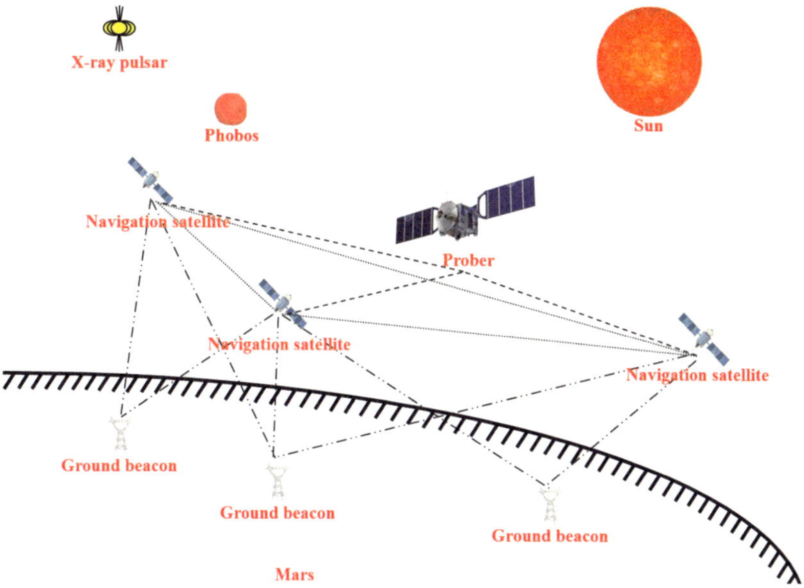

Figure 3.
The sketch map of the future development of the MGNSS.

the standard formulation issues, at present lots of international standards have been made for the satellite navigation system. For example, the International Organization for Standardization (ISO) has made a series of standards for the GPS, including its working channel, its coding and decoding method, and its service agreement. In future, some corresponding standards should also be made for the Mars Global Navigation Satellites System (MGNSS) [79]; **Figure 3** shows its imaginary representation.

5. Conclusion

In this article, the autonomous navigation for the Mars exploration is reviewed and summarized. First, the popular autonomous navigation techniques are presented. The inertial navigation, the celestial navigation, the visual navigation, and the integrated navigation methods are all introduced. Their advantages and the disadvantages are presented respectively. Second, the specific autonomous navigation for the Mars exploration is addressed. The corresponding navigation techniques are illustrated for different mission phases, including the launch phase, the cruise flight phase, the EDL phase, and the roving phase. Both the autonomous navigation and the nonautonomous navigation are compared. Third, the challenges and the future development trending of Mars exploration are summarized. Some new developed techniques are illustrated. This article can help the students and the researchers to know the autonomous navigation technology in the Mars exploration mission well.

Acknowledgements

We have used some pictures from the Internet including the photos and the cartoons. We express our thanks to the authors. This work was supported by the National Natural Science Foundation of China under Grant No. 61975011 and the Fundamental Research Fund for the China Central Universities of USTB under Grant No. FRF-BD-19-002A.

Conflict of interest

The authors declare no conflict of interest.

Author details

Haoting Liu
Beijing Engineering Research Center of Industrial Spectrum Imaging, School of Automation and Electrical Engineering, University of Science and Technology, Beijing, China

*Address all correspondence to: liuhaoting@ustb.edu.cn

References

[1] Zubrin RM, Baker DA. Mars direct: Humans to the red planet by 1999. Acta Astronautica. 1992;**26**:899-912. DOI: 10.1016/0094-5765(92)90130-B

[2] Haberle RM, Kahre MA, Hollingsworth JL, Montmessin F, Wilson RJ, Urata RA, et al. Documentation of the NASA/Ames legacy Mars global climate model: Simulation of the present seasonal water cycle. Icarus. 2019;**333**:130-164. DOI: 10.1016/j.icarus.2019.03.026

[3] Siddle AG, Mueller-Wodarg ICF, Ston SW, Yelle RV. Global characteristics of gravity waves in the upper atmosphere of Mars as measured by MAVEN/NGIMS. Icarus. 2019;**333**: 12-21. DOI: 10.1016/j.icarus.2019.05.021

[4] Zhao Y, Wang X, Li Q, Wang D, Cai Y. A high-accuracy autonomous navigation scheme for the Mars rover. Acta Astronautica. 2019;**154**:18-32. DOI: 10.1016/j.actaastro.2018.10.036

[5] Heldmann JL, Stoker CR, Gonzales A, McKay CP, Davila A, Glass BJ, et al. Red dragon drill missions to Mars. Acta Astronautica. 2017;**141**:79-88. DOI: 10.1016/j.actaastro.2017.10.002

[6] Conte D, Carlo MD, Ho K, Spencer DB, Vasile M. Earth-Mars transfers through moon distant retrograde orbits. Acta Astronautica. 2018;**143**:372-379. DOI: 10.1016/j.actaastro.2017.12.007

[7] Li S, Jiang X. Summary and enlightenment of GNC schemes for Mars entry, descent and landing. Journal of Astronautics. 2016;**37**:499-511. DOI: 10.3873/j.issn.1000-1328.2016.05.001

[8] Giorgi G, Schmidt TD, Trainotti C, Mata-Calvo R, Fuchs C, Hoque MM, et al. Advanced technologies for satellite navigation and geodesy. Advances in Space Research. 2019;**64**:1256-1273. DOI: 10.1016/j.asr.2019.06.010

[9] Yu M, Li S, Wang S, Huang X. Single crater-aided inertial navigation for autonomous asteroid landing. Advances in Space Research. 2019;**63**:1085-1099. DOI: 10.1016/j.asr.2018.09.035

[10] Wang S, Cui P, Gao A, Yu Z, Cao M. Absolute navigation for Mars final approach using relative measurements of X-ray pulsars and Mars orbiter. Acta Astronautica. 2017;**138**:68-78. DOI: 10.1016/j.actaastro.2017.05.017

[11] Arai K, Takamura H, Inoue H, Ono M, Adachi S. Fast vision-based localization for a Mars airplane. In: Proceedings of the SICE Annual Conference (SCIE '14); 9-12 September 2014; Sapporo, Japan. New York: IEEE; 2014. pp. 1449-1455

[12] Yang H, Fu H, Wang Z, Zhang Y. Autonomous navigation for Mars final approach using X-ray pulsars and Mars network based on RKF. In: Proceedings of the European Navigation Conference (ENC '19); 9-12 April 2019; Warsaw, Poland. New York: IEEE; 2019. pp. 1-5

[13] Liu J, Fang J, Liu G, Wu J. Solar flare TDOA navigation method using direct and reflected light for Mars exploration. IEEE Transactions on Aerospace and Electronic Systems. 2017;**53**:2469-2484. DOI: 10.1109/TAES.2017.2700958

[14] Jiang X, Tao T, Li S. Aerostats/orbiters radiometric measurement based integrated navigation for Mars descent and landing. In: Proceedings of the Chinese Control Conference (CCC '15); 28-30 July 2015; Hangzhou, China. New York: IEEE; 2015. pp. 5473-5478

[15] Li Z, Wang Y, Zheng W, Lian S. Spacecraft autonomous navigation via inter-spacecraft relaying

communication for Mars prober. In: Proceedings of the International Conference on Information Fusion (FUSION '17); 10-13 July 2017; Xi'an China. New York: IEEE; 2017. pp. 1-5

[16] Lou T, Liu J, Jin P. An innovative integrated navigation based on norm-constrained UKF during Mars powered descent. In: Proceedings of the IEEE Chinese Guidance, Navigation and Control Conference (CGNCC '16); 12-14 August, 2016; Nanjing, China. New York: IEEE; 2016. pp. 663-668

[17] Konybaev NB, Ibadulla SI, Diveev AI. Evolutional method for creating artificial intelligence of robotic technical systems. Procedia Computer Science. 2019;**150**:709-715. DOI: 10.1016/j.procs.2019.02.018

[18] Yu Z, Cui P, Crassidis JL. Design and optimization of navigation and guidance techniques for Mars pinpoint landing: Review and prospect. Progress in Aerospace Science. 2017;**94**:82-94. DOI: 10.1016/j.paerosci.2017.08.002

[19] Ning X, Gui M, Zhang J, Fang J. A dimension reduced INS/VNS integrated navigation method for planetary rovers. Chinese Journal of Aeronautics. 2016;**29**:1695-1708. DOI: 10.1016/j.cja.2016.10.009

[20] Ning X, Gui M, Fang J, Liu G. Differential X-ray pulsar aided celestial navigation for Mars exploration. Aerospace Science and Technology. 2017;**62**:36-45. DOI: 10.1016/j.ast.2016.10.032

[21] Balakalyani G, Jagadeesh G. An accelerometer balance for aerodynamic force measurements over hypervelocity ballistic models in shock tunnel. Measurement. 2019;**136**:636-646. DOI: 10.1016/j.measurement.2018.12.099

[22] Wang X, Xiao L. Gyroscope-reduced inertial navigation system for flight vehicle motion estimation. Advances in Space Research. 2017;**59**:413-424. DOI: 10.1016/j.asr.2016.09.001

[23] Christophe B, Boulanger D, Foulon B, Huynh P-A, Lebat V, Liorzou F, et al. A new generation of ultra-sensitive electrostatic accelerometers for GRACE follow-on and towards the next generation gravity missions. Acta Astronautica. 2015;**117**: 1-7. DOI: 10.1016/j.actaastro.2015.06.021

[24] Demenkov NP, Tran DM. Fuzzy description hemispherical resonator gyro error. Procedia Computer Science. 2019;**150**:88-94. DOI: 10.1016/j.procs.2019.02.019

[25] Liu H, Shi D, Hou X, Yan B, Wang W. Manufacture process quality control of interferometric fiber optic gyroscope using analyses of multi-type assembly and test data. International Journal of Computer Integrated Manufacturing. 2018;**11**:1124-1140. DOI: 10.1080/0951192X.2018.1509128

[26] Solouk MR, Shojaeefard MH, Dahmardeh M. Parametric topology optimization of a MEMS gyroscope for automotive applications. Mechanical Systems and Signal Processing. 2019;**128**:389-404. DOI: 10.1016/j.ymssp.2019.03.049

[27] Tang Q, Wang X. Autonomous navigation based on Deimos light-of-sight for Mars approaching phase. Chinese Space Science and Technology. 2019;**39**:38-46. DOI: 10.16708/j.cnki.1000-758X.2019.0048

[28] Wang X, Zheng R, Wu Y, Sun J, Zhong J, Wang L, et al. Study on the method of precision adjustment of star sensor. Nanotechnology and Precision Engineering. 2018;**1**:248-257. DOI: 10.1016/j.npe.2018.12.001

[29] Wu W, Ma X, Ning X. Autonomous navigation method with high accuracy for cruise phase of Mars probe. SCIENCE CHINA Information

Sciences. 2012;**42**:936-948. DOI: 10.1360/112011-1366

[30] Liu J, Ma J, Tian J. Integrated X-ray pulsar and Doppler shift navigation. Journal of Astronautics. 2010;**31**:1552-1557. DOI: 10.3873 j.issn.1000-1328.2010.06.007

[31] Liu J, Fang J, Ma X, Kang Z, Wu J. X-ray pulsar/starlight Doppler integrated navigation for formation flight with ephemerides errors. IEEE Aerospace and Electronic Systems Magazine. 2015;**30**:30-39. DOI: 10.1109/MAES.2014.140074

[32] Yang H, Wang Z, Fu H, Zhang Y. Integrated navigation for Mars final approach based on Doppler radar and X-ray pulsars with atomic clock error. Acta Astronautica. 2019;**159**:308-318. DOI: 10.1016/j.actaastro.2019.03.055

[33] Takenaka T, Toyota S, Kuroda H, Kobayashi M, Kumagai T, Mori K, et al. Freehand technique of an electromagnetic navigation system emitter to avoid interference caused by metal neurosurgical instruments. World Neurosurgery. 2018;**118**:143-147. DOI: 10.1016/j.wneu.2018.07.071

[34] Vigneron AC, Ruiter AHJD, Burlton BV, Soh WKH. Nonlinear filtering for autonomous navigation of spacecraft in highly elliptical orbit. Acta Astronautica. 2016;**126**:138-149. DOI: 10.1016/j.actaastro.2016.03.035

[35] Zheng N, Liu H. Robotic precise positioning and interaction based on laser radar. In: Proceedings of the IEEE International Conference on Robotics and Biomimetics (ROBIO '18); 12-15 December 2018; Kuala Lumpur, Malaysia. New York: IEEE; 2018. pp. 206-211

[36] Shih Y. Quantum imaging. IEEE Journal of Selected Topics in Quantum Electronics. 2007;**13**:1016-1030. DOI: 10.1109/JSTQE.2007.902724

[37] Guan X, Wang X. Review of vision-based navigation technique. Aero Weaponry. 2014;**5**:3-14. DOI: 10.19297/j.cnki.41-1228/tj.2014.05.001

[38] Xu H, Xu J, Xu W. Survey of 3D modeling using depth cameras. Virtual Reality & Intelligent Hardware. 2019;**1**:483-499. DOI: 10.1016/j.vrih.2019.09.003

[39] Wang D, Xu C, Huang X. Overview of autonomous navigation based on sequential images for planetary landing. Journal of Harbin Institute of Technology. 2016;**48**:1-12. DOI: 10.11918 /j.issn.0367-6234.2016.04.001

[40] Nguyen D-V, Kuhnert L, Kuhnert KD. Structure overview of vegetation detection. A novel approach for efficient vegetation detection using an active lighting system. Robotics and Autonomous Systems. 2012;**60**:498-508. DOI: 10.1016/j.robot.2011.11.012

[41] Liu H, Lu H, Zhang Y. Image enhancement for outdoor long-range surveillance using IQ-leaning multiscale Retinex. IET Image Processing. 2017;**11**:786-795. DOI: 10.1049/iet-ipr.2016.0972

[42] Li W, Zhang L, Sun F, Yang L, Chen M, Li Y. Alignment calibration of IMU and Doppler sensors for precision INS/DVL integrated navigation. Optik. 2015;**126**:3872-3876. DOI: 10.1016/j.ijleo.2015.07.187

[43] Gu L, Jiang X, Li S, Li W. Optical/radio/pulsars integrated navigation for Mars orbiter. Advances in Space Research. 2019;**63**:512-525. DOI: 10.1016/j.asr.2018.09.005

[44] Li J, Wang D. Information-fusion-integrated navigation for satellite around Mars. In: Proceedings of the Chinese Control Conference (CCC '16); 27-29 July 2016; Chengdu, China. New York: IEEE; 2016. pp. 5441-5445

[45] Sun Y, Deng Y, Duan H, Xu X. Bionic visual close-range navigation control system for the docking stage of probe-and-drogue autonomous aerial refueling. Aerospace Science and Technology. 2019;**91**:136-149. DOI: 10.1016/j.ast.2019.05.005

[46] Jiang X, Li S, Huang X. Radio/FADS/IMU integrated navigation for Mars entry. Advances in Space Research. 2018;**61**:1342-1358. DOI: 10.1016/j.asr.2017.12.010

[47] Dhara S, Burleigh S, Datta R, Ghose S. CGR-BF: An efficient contact utilization scheme for predictable deep space delay tolerant network. Acta Astronautica. 2018;**151**:401-411. DOI: 10.1016/j.actaastro.2018.06.023

[48] Lee B-S, Hwang W, Kim H-Y, Kim J. Design and implementation of the flight dynamics system for CMOS satellite mission operations. Acta Astronautica. 2011;**68**:1292-1306. DOI: 10.1016/j.actaastro.2010.09.002

[49] Wu Y, Yao J, Qu X. An adaptive reentry guidance method considering the influence of blackout zone. Acta Astronautica. 2018;**142**:253-264. DOI: 10.1016/j.actaastro.2017.10.041

[50] Krishnamurthy A, Villasenor J, Seager S, Ricker G, Vanderspek R. Precision characterization of the TESS CCD detectors: Quantum efficiency, charge blooming and undershoot effects. Acta Astronautica. 2019;**160**:46-55. DOI: 10.1016/j.actaastro.2019.04.016

[51] Kubota T, Sato T, Ejiri R. Experimental study on visual navigation for exploration robot. IFAC Proceedings Volumes. 2010;**43**:169-174. DOI: 10.3182/20100906-3-IT-2019.00031

[52] Zhang Q, Chen MES, Li B. A visual navigation algorithm for paddy field weeding robot based on image understanding. Computers and Electronics in Agriculture. 2017;**143**:66-78. DOI: 10.1016/j.compag.2017.09.008

[53] Zhang Y, Xiao M, Wang Z, Fu H, Wu Y. Robust three-stage unscented Kalman filter for Mars entry phase navigation. Information Fusion. 2019;**51**:67-75. DOI: 10.1016/j.inffus.2018.11.003

[54] Blok PM, Boheemen KV, Evert FKV, IJsselmuiden J, Kim G-H. Robot navigation in orchards with localization based on particle filter and Kalman filter. Computers and Electronics in Agriculture. 2019;**157**:261-269. DOI: 10.1016/j.compag.2018.12.046

[55] Ning X, Huang P. Adaptive points range square-root cubature Kalman filter for Mars approach navigation. In: Proceedings of the Chinese Control Conference (CCC '14); 28-30 July, 2014; Nanjing, China. New York: IEEE; 2014. pp. 903-908

[56] Wang W, Wei X, Li J, Wang G. Noise suppression algorithm of short-wave infrared star image for daytime star sensor. Infrared Physics & Technology. 2017;**85**:382-394. DOI: 10.1016/j.infrared.2017.08.002

[57] Cui P, Gao X, Zhu S, Shao W. Visual navigation based on curve matching for planetary landing in unknown environments. Acta Astronautica. 2020;**170**:261-274. DOI: 10.1016/j.actaastro.2020.01.023

[58] Chiang KW, Tsai GJ, Chang HW, Joly C, El-Sheimy N. Seamless navigation and mapping using an INS/GNSS/grid-based SLAM semi-tightly coupled integration scheme. Information Fusion. 2019;**50**:181-196. DOI: 10.1016/j.inffus.2019.01.004

[59] Mil S, Piantanakulchai M. Modified Bayesian data fusion model for travel time estimation considering spurious data and traffic conditions. Applied

Soft Computing. 2018;**72**:65-78. DOI: 10.1016/j.asoc.2018.06.046

[60] Kampa K, Principe JC, Slatton KC. Dynamic factor graphs: A novel framework for multiple features data fusion. In: Proceedings of the IEEE International Conference on Acoustics, Speech and Signal Processing (ICASSP '10); 14-19 March, 2010; Dallas, TX, USA. New York: IEEE; 2010. pp. 2106-2109

[61] Meng T, Jing X, Yan Z, Pedrycz W. A survey on machine learning for data fusion. Information Fusion. 2020;**57**:115-129. DOI: 10.1016/j.inffus.2019.12.001

[62] Zhou S, Hu X, Liu L, He F, Tang C, Pang J. Status of satellite orbit determination and time synchronization technology for global navigation satellite systems. Chinese Astronomy and Astrophysics. 2019;**43**:479-492. DOI: 10.1016/j.chinastron.2019.11.003

[63] Re E, Cintio AD, Busca G, Giunta D, Sanchez M. Novel time synchronization techniques for deep space probes. In: Proceedings of the European Navigation Conference (ENC '09); 20-24 April, 2009; Besancon, France. New York: IEEE; 2009. pp. 205-210

[64] Li Z, Wang Y, Zheng W. Observability analysis of autonomous navigation using inter-satellite range: An orbital dynamics perspective. Acta Astronautica. 2020;**170**:577-585. DOI: 10.1016/j.actaastro.2020.02.023

[65] Ou Y, Zhang H. Mars final approach navigation using ground beacons and orbiters: An information propagation perspective. Acta Astronautica. 2017;**138**:490-500. DOI: 10.1016/j.actaastro.2017.06.010

[66] Guan RP, Ristic B, Wang L, Evans R. Monte Carlo localization of a mobile robot using a Doppler-azimuth radar. Automatica. 2018;**97**:161-166. DOI: 10.1016/j.automatica.2018.08.012

[67] Jiang X, Duan F, Tian H, Wei X. Optimization of reliability centered predictive maintenance scheme for inertial navigation system. Reliability Engineering and System Safety. 2015;**140**:208-217. DOI: 10.1016/j.ress.2015.04.003

[68] Andrade C, Martinez I, Castellote M, Zuloaga P. Some principles of service life calculation of reinforcements and in situ corrosion monitoring by sensors in the radioactive waste containers of El Cabril disposal (Spain). Journal of Nuclear Materials. 2006;**358**:82-95. DOI: 10.1016/j.jnucmat.2006.06.015

[69] Krobka NI. A new gyroscopic principle. New gyroscopic effects on cold atoms and on de Broglie waves, different from the Sagnac effect. Acta Astronautica. 2019;**163**:181-189. DOI: 10.1016/j.actaastro.2019.03.066

[70] Jiang P, Wang Z, Luo H. Techniques for measuring transverse relaxation time of xenon atoms in nuclear-magnetic-resonance gyroscopes and pump-light influence mechanism. Optik. 2017;**138**:341-348. DOI: 10.1016/j.ijleo.2017.03.043

[71] Song P, Ma J, Ma Z, Zhang S, Si C, Han G, et al. Research and development status of quantum navigation technology. Laser & Optoelectronics Progress. 2018;**55**:090003. DOI: 10.3788/LOP55.090003

[72] Wang Q, Zhou J. Simultaneous localization and mapping method for geomagnetic aided navigation. Optik. 2018;**171**:437-445. DOI: 10.1016/j.ijleo.2018.06.069

[73] Chen P, Han J, Lai Y, Sun X, Tan L. Autonomous navigation for middle and low orbit Mars

spacecraft based on gravity gradient measurement. Spacecraft Engineering. 2018;**27**:17-23. DOI: 10.3969/j.issn.1673-8748.2018.03.003

[74] Gao Z, Shi Q, Toshio F, Li C, Huang Q. An overview of biomimetic robots with animal behaviors. Neurocomputing. 2019;**332**:339-350. DOI: 10.1016/j.neucom.2018.12.071

[75] Gao S, Xue L, Zhong Y, Gu C. Random weighting method for estimation of error characteristics in SINS/GPS/SAR integrated navigation system. Aerospace Science and Technology. 2015;**46**:22-29. DOI: 10.1016/j.ast.2015.06.029

[76] Mehmood H, Tripathi NK. Cascading artificial neural networks optimized by genetic algorithms and integrated with global navigation satellite system to offer accurate ubiquitous positioning in urban environment. Computers, Environment and Urban Systems. 2013;**37**:35-44. DOI: 10.1016/j.compenvurbsys.2012.04.004

[77] Patle BK, Babu LG, Pandey A, Parhi DRK, Jagadeesh A. A review: On path planning strategies for navigation of mobile robot. Defence Technology. 2019;**15**:582-606. DOI: 10.1016/j.dt.2019.04.011

[78] Kumar PB, Sahu C, Parhi DR. A hybridized regression-adaptive ant colony optimization approach for navigation of humanoids in a cluttered environment. Applied Soft Computing. 2018;**68**:565-585. DOI: 10.1016/j.asoc.2018.04.023

[79] Han T, Shi X, Jian N, Fung L, Ping J. A preliminary designation of constellation distribution for Mars GNSS. In: Proceedings of the Asia-Pacific Conference on Communications (APCC '09); 8-10 October; Shanghai, China. New York: IEEE; 2009. pp. 306-309

Chapter 3

Aerocapture, Aerobraking, and Entry for Robotic and Human Mars Missions

Ye Lu

Abstract

This chapter provides an overview of the aeroassist technologies and performances for Mars missions. We review the current state-of-the-art aeroassist technologies for Mars explorations, including aerocapture, aerobraking, and entry. Then we present a parametric analysis considering key design parameters such as interplanetary trajectory and vehicle design parameters (lift-to-drag ratio, ballistic coefficient, peak g-load, peak heat rate, and total heat load) for aerocapture, aerobraking, and entry. A new perspective on a rapid aerobraking concept will be provided. The analysis will include first-order estimates for thermal loading, thermal protection systems material selection, and vehicle design. Results and discussion focus on both robotic missions and human missions as landed assets and orbiters.

Keywords: aerocapture, aerobraking, entry, robotic mission, human mission

1. Introduction

Aeroassist maneuvers are a family of maneuvers that use aerodynamic forces to change a spacecraft orbit and they include atmospheric entry, aerocapture, aerobraking, and aerogravity-assist. Atmospheric entry is used for in situ explorations, both for robotic and human missions. Atmospheric entry at Mars has been attempted many times by multiple space agencies. Entry at Mars was considered a challenging task mainly due to the unique atmospheric structure [1]. The atmosphere is substantial that aerothermodynamic heating is a consideration, yet the atmosphere is very thin that the aerodynamic drag is barely enough for entry vehicles to decelerate to a velocity at high altitude to safely initiate the final descent stage for a soft landing (i.e., parachute or retro-propulsion).

The concept of using atmosphere to change orbit can be traced back to the earliest publication by London in 1961 [2], which later evolved intro three main categories—aerobraking, aerocapture, and aerogravity-assist. Aerobraking is a maneuver where spacecraft uses atmospheric drag to reduce its orbital period, and it can be used for orbit transfer vehicles from GEO to LEO, or after initial orbit insertion for planetary missions. In the context of Mars missions, aerobraking maneuver is considered free in terms of system requirement because no additional system/mass is needed to perform the maneuver. All the prior aerobraking spacecraft use solar panel as the drag device to decelerate. However, aerobraking maneuver is not free in terms of operational cost. Due to the long duration of aerobraking maneuver—on the order of months, constant ground operation is required in the past for aerobraking maneuver, which requires hours of staffs and dedicated time with the Deep Space Network (DSN) for position tracking [3].

Aerocapture is an orbit insertion maneuver. Upon first approaching a planet upon hyperbolic trajectory, the spacecraft passes the body's atmosphere once to decelerate and achieve a captured orbit after the single pass. Aerocapture maneuver has been studied in the literature but has never been tested or demonstrated in flight. Aerocapture at Mars is considered side-by-side with aerobraking or direct propulsive orbit insertion.

Aerogravity-assist is a maneuver for interplanetary transfer and most often considered for fast transfer time to the outer solar system for which Mars can be used as a destination to perform aerogravity-assist maneuver, therefore aerogravity-assist will not be discussed in detail in this chapter.

The structure of the chapter is as follows: in Section 2 we discuss the mathematical models and summarize the key parameters for aeroassist maneuvers and vehicles. In Section 3, aerobraking technology is presented along with a new perspective on aerobraking at Mars. In Section 4, we discuss Mars entry technology and the system performance and requirements for future missions. In Section 5, we discuss the performance and system requirements for aerocapture.

2. Mathematical models

2.1 Equations of motion

Assuming a nonrotating body in the body-centered and body-fixed reference frame, the equations of motions for entry, aerocapture, and aerobraking are as follows [4]:

$$\dot{\theta} = \frac{V \cos\gamma \cos\chi}{r \cos\phi} \tag{1}$$

$$\dot{\phi} = -\frac{V \cos\gamma \sin\chi}{r} \tag{2}$$

$$\dot{r} = V \sin\gamma \tag{3}$$

$$\dot{V} = -\frac{q}{\beta} - g \sin\gamma \tag{4}$$

$$\dot{\gamma} = -\frac{q(L/D)}{V\beta} \cos\sigma + \left(\frac{V}{r} - \frac{g}{V}\right) \cos\gamma \tag{5}$$

$$\dot{\chi} = \frac{q(L/D)}{V\beta} \frac{\cos\sigma}{\cos\gamma} + \frac{V}{r} \cos\gamma \sin\chi \tan\phi + \frac{g}{V} \frac{\cos\chi}{\cos\gamma} \tag{6}$$

where θ and ϕ are the longitude and latitude in a spherical surface model; r the radial distance from the center; χ is the heading angle measured clockwise from the direction of local parallel; V is the velocity of the vehicle; γ is the flight path angle (positive above local horizon); σ is the bank angle, which is the rotation angle about the relative velocity vector; $\beta = m/(C_D A)$ is the ballistic coefficient where m is the vehicle mass and C_D is the aerodynamic drag coefficient; and $q = (1/2)\rho V^2$ is the dynamic pressure, where ρ is the density of the atmosphere, and L and D are the lift and drag forces respectively and are defined as:

$$L = \frac{1}{2}\rho V^2 A C_L, \qquad D = \frac{1}{2}\rho V^2 A C_D \tag{7}$$

where C_L is the aerodynamic lift coefficient, and A is the aerodynamic reference area of the vehicle. The angle of attack, α, affects the value of C_L and C_D, and is assumed constant as the trim angle of attack. $g = \mu/r^2$ is the radial component of the gravitational acceleration.

These equations of motion are used throughout this chapter for numerical analysis of trajectories during atmospheric fly-through or entry.

2.2 Aerothermodynamic heating

As the spacecraft flies through the atmosphere at hypersonic speed, the aerothermodynamic heating can be substantial for entry and aerocapture. Aerothermodynamic heating consists of mainly two types of dominating heat transfers: convective and radiative. Sutton and Graves [5] developed an empirical relation for convective heat rate with an arbitrary gas mixture:

$$\dot{q}_c = k(\rho/R_n)^{0.5} V^3 \tag{8}$$

where R_n is the nose radius in m. $\dot{q}_c$ has the unit of W/cm^2 and k has a value of 1.8980×10^{-8}, ρ is the atmospheric density. The empirical relation for radiative heating rate, $\dot{q}_r$ is W/cm^2 follows [6]:

$$\dot{q}_r = 2.35 \times 10^4 r_n^{0.526} \rho^{1.19} f(V) \tag{9}$$

where $f(V)$ is provided for velocities between 6000 m/s and 9000 m/s and is shown in **Figure 1**. At low speed, radiative heating becomes insignificant due to $f(V)$ approaching zero.

2.3 Vehicle designs

Thermal protection system (TPS) is an important vehicle component for all aeroassist maneuvers to protect the spacecraft from the heat generated during the atmospheric pass. Entry and aerocapture can share the vehicle designs as both maneuvers result in very high heat, which warrants TPS. However, aerobraking maneuvers have been achieved without a dedicated TPS and solar panels had been used as the drag device to reduce orbital period; therefore, the following discussion

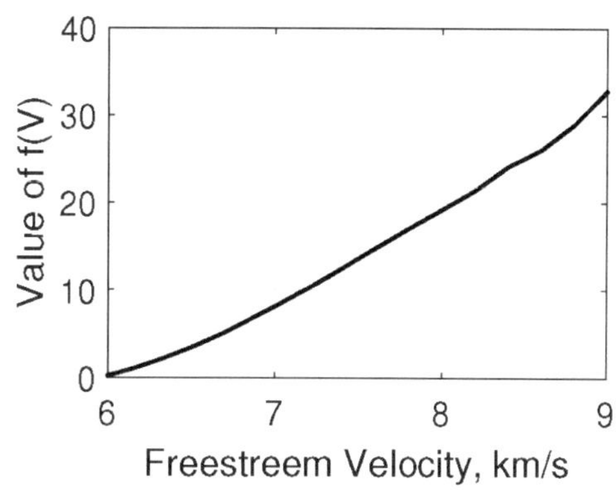

Figure 1.
Coefficient value $f(V)$ as a function of velocity for radiative heating relation [6].

on vehicle designs is applicable to aerocapture and entry. Heritage blunt-body rigid aeroshell designs have been proven for both robotic and manned missions. Most robotic missions used ballistic entry vehicles, which have no active guidance or control (e.g., Mars Pathfinder, Mars Exploration Rovers (MER), and Mars Phoenix), whereas lifting body entry vehicles are used for manned missions and some Mars missions; for example, Mars Science Laboratory used a lifting vehicle in order to meet the landing accuracy requirement and Apollo entry capsules met the safe g-load limit acceptable to humans.

With lifting vehicle design, the guidance and control actively modulate the direction of the lift vector thus to control the trajectory, which is also called bank modulation. The lifting vehicles are typically designed with a nominal L/D with center-of-mass offset or asymmetric heatshield, and are also equipped with thrusters to control the orientation of the lifting vector. **Table 1** lists some lifting body entry vehicles. Vehicle design with spherical section and sphere-cones are most popular and have been used for all entry missions and they can provide L/D of more than 0.3. These rigid body aeroshells are shown feasible for both entry and aerocapture missions for various planetary bodies [8–10].

Adaptable Deployable Entry and Placement Technology (ADEPT) [11] and Hypersonic Inflatable Aerodynamic Decelerator (HIAD) [12] are deployable entry systems that are currently being developed. Both ADEPT and HIAD are applicable to a range of mission sizes from small satellites to larger payloads. Ellipsled vehicle design, or mid-L/D vehicle, has been proposed as a means to increase vehicle control authority (e.g., for ice giants missions [8]) or deliver higher payload mass, such as for human Mars architectures. Starbody waverider can achieve higher nominal L/D (>5.0) than other designs and is mostly useful for interplanetary transfer maneuvers such as aerogravity-assist; therefore, it is only referenced here for comparison [7, 13]. Higher L/D vehicles are available, but in the context of Mars missions, they have very limited applications.

Drag modulation, in addition to bank modulation, is another design that provides the vehicle control authority. Drag modulation uses ballistic vehicles but with additional drag skirt that can be modified to change the vehicle's ballistic coefficient, thus achieving trajectory control in the atmosphere. Such vehicles would require a large ratio for the designed low and high values of ballistic coefficients. Drag modulations can be used for both entry and aerocapture at Mars [14, 15].

Angle-of-attack modulation has also been investigated for Mars entry [16] and shown feasible for aerocapture missions [17]. Direct force control is yet another control mode for entry vehicles, which uses active flaps to create moments and controls the angle-of-attack and side slip angles [18]. Both of the control modes

Vehicle design	Planet	Mission (year)	Entry mass, kg	$(L/D)_{trim}$
Spherical section	Earth	Apollo (1960s)	5560	>0.30
Sphere-cone	Mars	Viking I and II (1976)	576	0.18
Sphere-cone	Mars	MSL (2012)	3380	0.24
ADEPT/HIAD	—	—	Variable	~0.2
Ellipsled	Neptune	—	~6000	0.6–0.8
—	Earth	Space Shuttle	~100,000	1 (hypersonic)
Starbody Waverider	—	Aeroassist	~100,00	~2 [7]

Table 1.
Performance and spacecraft parameters for past Venus and Mars missions using aerobraking.

require a high accuracy in hypersonic flow modeling and the uncertainties at hypersonic speed can be very difficult to predict; therefore, they have mostly been studied in paper and has not been implemented in missions.

Magnetohydrodynamics flow control is another means to actively control the trajectory, which uses the Lorentz force (i.e., the interaction between the plasma field from the hypersonic entry and magnetic field) [19]. It has been shown useful for entry trajectory control and similar is applicable for aerocapture trajectory control. Last but not the least is applying propulsion during aerocapture maneuver to create propulsive "lift" force in order to achieve the necessary trajectory control [20].

3. Aerobraking at Mars

Aerobraking maneuver was first successfully demonstrated at Venus with Magellan mission in 1993 after completing its prime mission. Magellan used aerobraking maneuver to reduce its orbital period from 3.23 h to 1.57 h. Following the Magellan's success, three Mars missions have used aerobraking as an enabling technology to reduce the propellant requirement to enter the target science orbits. The three missions are Mars Global Surveyor (MGS), Mars Odyssey, and Mars Reconnaissance Orbiter (MRO), which were launched in 1996, 2001, and 2005 respectively. **Table 2** summarizes the spacecraft parameters and aerobraking performances for selected aerobraking missions. As shown, the fuel mass saving for all three Mars missions are all over 1000 m/s, which is very significant compared with the launch mass. According to the rocket equation, the propellant mass follows an exponential relation to the required ΔV, the amount of fuel savings can be considered enabling for MGS and Odyssey.

Aerobraking operation includes mainly three phases—walk-in phase, main phase, and walk-out phase. The walk-in phase follows the initial Mars orbit insertion, and reduces the periapsis altitude within Mars atmosphere. During the main phase of aerobraking, the spacecraft uses atmospheric drag to reduce the energy and apoapsis altitude. Past missions have used the solar panel as the main drag device. The

Spacecraft	**Magellan**	**MGS**	**Odyssey**	**MRO**
Destination	Venus	Mars	Mars	Mars
Launch mass, kg	360	1060	725	2180
Propellant mass, kg	2414	385	348.7	1149
Payload mass, kg	154	78	44.5	139
AB ΔV saving, m/s	1220	1220	1090	1190
AB fuel saving, kg	490	330	320	580
AB duration	70 days	6 months	2.5 months	5 months
Period before AB, h	3.2	45	18	34
Period after AB, h	1.6	1.9	2	1.9
AB periapsis range, km	171.3–196.9	100–149	107–119	97–110
Dynamic pressure, N/m^2	0.2–0.3	0.6[a]	0.2–0.3	—
Heat rate, W/cm^2	—	—	—	0.75–1.6

[a]*Dynamic pressure is reduced to 0.2 N/m^2 after the failure of solar panel hinge.*

Table 2.
Performance and spacecraft parameters for past Venus and Mars missions using aerobraking.

dynamic pressure, heat rate, or temperature on the solar panel are monitored to ensure the integrity of the solar panel. During the main phase, constant ground operations are needed in order to actively control the periapsis altitude via minor apoapsis burns so that the solar panel will not overheat. Over a course of months and after hundreds of atmospheric passes, the target apoapsis altitude will be achieved. The walk-out phase is simply a series of impulsive burns at the apoapsis to raise the periapsis altitude out of the atmosphere and to the target science orbit.

3.1 Aerobraking maneuver constraints

The primary constraints of aerobraking maneuver are the structural load and maximum temperature of the solar panel (or drag panel in general). As the spacecraft flies through the upper atmosphere, aerodynamic drags generate heat and increase the temperature of the solar panel. In addition, the drag forces exert structural load on the vehicle, in particular the hinges where the solar panel is connected to the main structure. Most spacecraft use deployable solar panels due to the volume restriction during launch; thus, the connector hinges of the solar panel are usually deployable. One measure of the structural load is the dynamic pressure at periapsis, which is also the maximum dynamic pressure. It is also found that maximum temperature on the solar panel is correlated with the dynamic pressure at periapsis. **Table 2** summarizes the constraints for dynamic pressure and heat rate of the previous aerobraking missions. Magellan, MGS, and Mars Odyssey all used dynamic pressure at periapsis as the constraint metric while MRO used heat rate as the measure of constraint. These values have been flight-tested and proven to be acceptable using only solar panels as the drag device.

3.2 Aerobraking with dedicated structure

Aerobraking maneuver has been considered "free" in terms of the mass budget for spacecraft and it has saved thousands of kilograms of propellant for the past missions as shown in **Table 2**. However, aerobraking maneuvers usually take months. It may be acceptable for robotics missions, but for human missions, spending several more months to perform aerobraking to save propellant mass may be prohibitive.

A novel way for faster aerobraking maneuver is by diving deeper in the atmosphere to achieve more deceleration from a single atmospheric pass, thus reducing the time required to perform the aerobraking maneuver. Dedicated aerobraking hardware using deployable structure and membrane has been investigated in Ref. [21] but with a similar heat rate and dynamic pressure as from the solar panels. With a lightweight deployable structure, the effective drag area is increased and higher deceleration can be achieved from every atmospheric pass.

Ballistic coefficient β in Eqs. (1)–(6) is an important parameter that relates the vehicle mass and the effective aerodynamic drag area. With all things equal, a larger drag area results in a lower ballistic coefficient while a heavier vehicle results in a higher ballistic coefficient. For robotic missions, the ballistic coefficients are usually in the range of 50–150 kg/m^2, whereas for human missions, high ballistic coefficients ($\sim$1000 kg/m^2) can be expected.

3.3 Aerobraking design trade

To evaluate the performance of aerobraking maneuver, a numerical analysis is conducted to obtain the key metrics, such as the dynamic pressure, peak heat rate, and effective ΔV. A range of vacant periapsis altitudes are assessed for the entire range of entry velocities, 3.49–4.93 km/s (i.e., velocity of a 129-km circular orbit and the escape velocity at entry altitude of 129 km). Different entry velocities and vacant periapsis altitudes result in different dynamic pressure, peak heat rate, and effective ΔV, which are shown as contours in **Figure 2**. It is important to note that the lower left area where the contours end corresponds to crash trajectories; that is, the vacant periapsis altitude is too low and the vehicle will enter the atmosphere entirely. For aerobraking, that is the restricting region.

While **Figure 2** shows the general trends of the design parameters, it only contains limited information to design aerobraking maneuver with a dedicated hardware. One parameter that is of the least interest is the vacant periapsis altitude. After every atmospheric pass, a very small correction ΔV is executed at apoapsis point in order to adjust the vacant periapsis altitude. Thus, vacant periapsis altitude can be adjusted during flight. By replacing the vacant periapsis with ΔV, and plotting the same metrics, such as dynamic pressure and peak heat rate on the chart, we obtain **Figure 3**. Note that orbital periods for each entry velocity and ΔV values are also plotted for reference. At a specific entry velocity, each ΔV value corresponds to only one vacant periapsis altitude. Therefore, vacant periapsis altitude is embedded in ΔV and has not been lost.

From **Figure 3**, we can identify the relations between the constraints (i.e., dynamic pressure or heat rate) with the orbital maneuvers. Using a dynamic pressure limit of 0.5 N/m^2, we can track the dynamic pressure contour of 0.5 N/m^2 from the right to the left. As the orbital period decreases, with the same dynamic pressure limit, more ΔV can be achieved at lower velocity. Starting an orbital period of 240 h (10 days), with a limit of 0.5 N/m^2, the initial ΔV per atmospheric pass is about 0.3 m/s per atmospheric pass. If the orbital period is reduced from 240 h to 24 h (i.e., reducing velocity from 4.88 km/s to 4.71 km/s), it will take approximately 450 atmospheric passes, which is equivalently more than 6 years in duration.

In order to show the effects of ballistic coefficients on aerobraking performance, **Figures 4** and **5** present the design parameters for ballistic coefficients of 50 kg/m^2 and 900 kg/m^2 respectively. By comparison, vehicles with lower ballistic

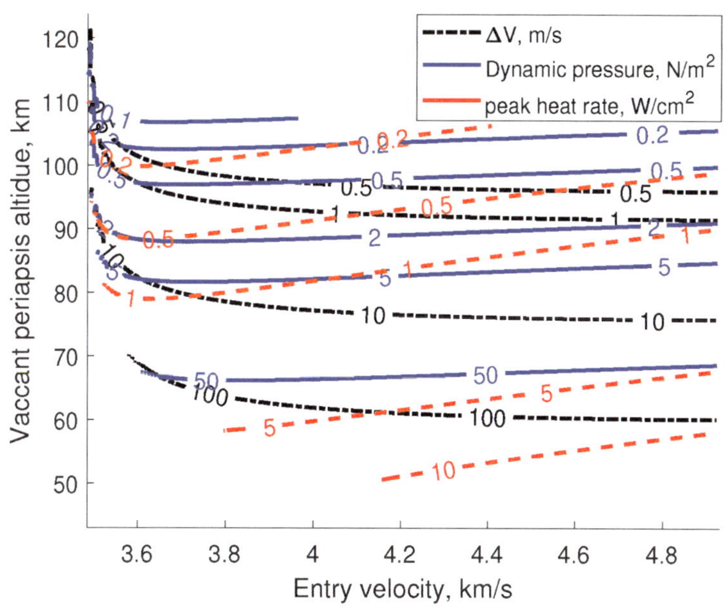

Figure 2.
Performance and design parameters for aerobraking maneuver for vehicle ballistic coefficient of 200 kg/m².

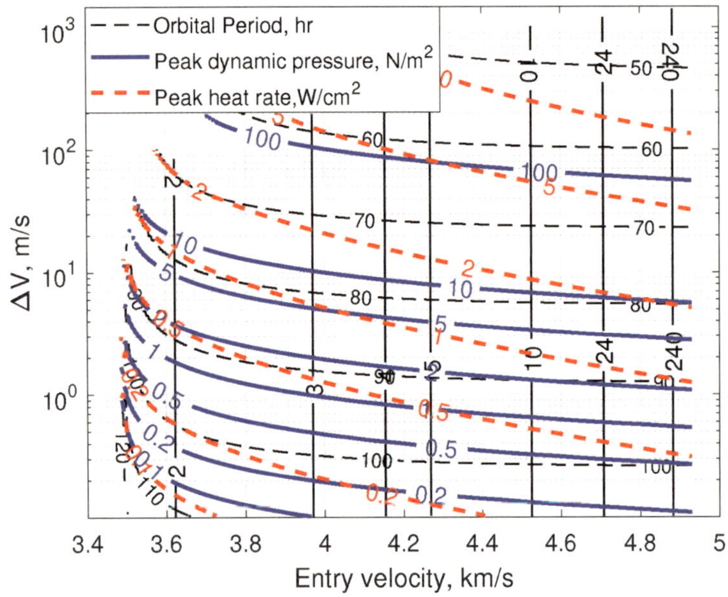

Figure 3.
Performance and design parameters for aerobraking maneuver for vehicle ballistic coefficient of 200 kg/m².

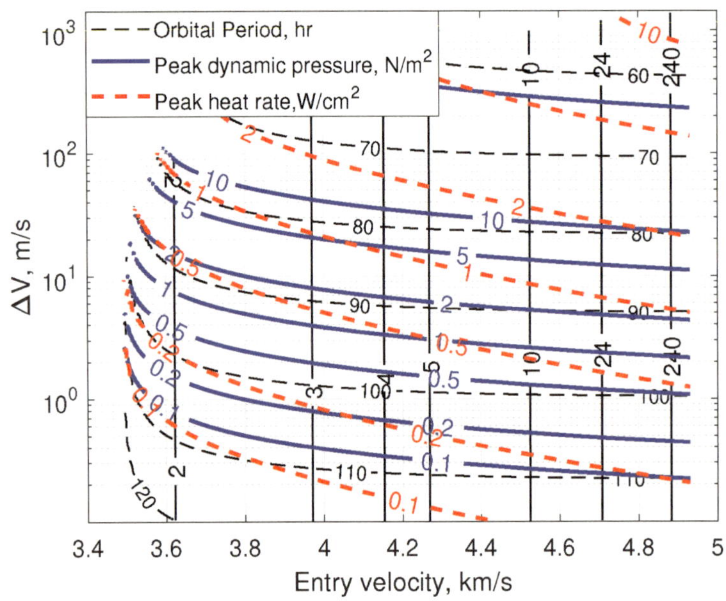

Figure 4.
Performance and design parameters for aerobraking maneuver for vehicle ballistic coefficient of 50 kg/m².

coefficients can decelerate at a higher rate with the same constraints. Considering the same limit of 0.5 N/m^2 for ballistic coefficient of 50 kg/m^2, the ΔV at 240-h period orbit is about 1.3 m/s. A ballistic coefficient of 50 kg/m^2 is roughly the value for the past aerobraking missions, and the duration for aerobraking maneuver to reduce the orbital period from 34 h to 1.9 h (as for MRO) is about 5 months.

For potential human missions using aerobraking, that is, assuming a ballistic coefficient of 900 kg/m^2, the vehicle needs to take a more aggressive approach in terms of the constraints in order to achieve the desired orbit in the same amount of

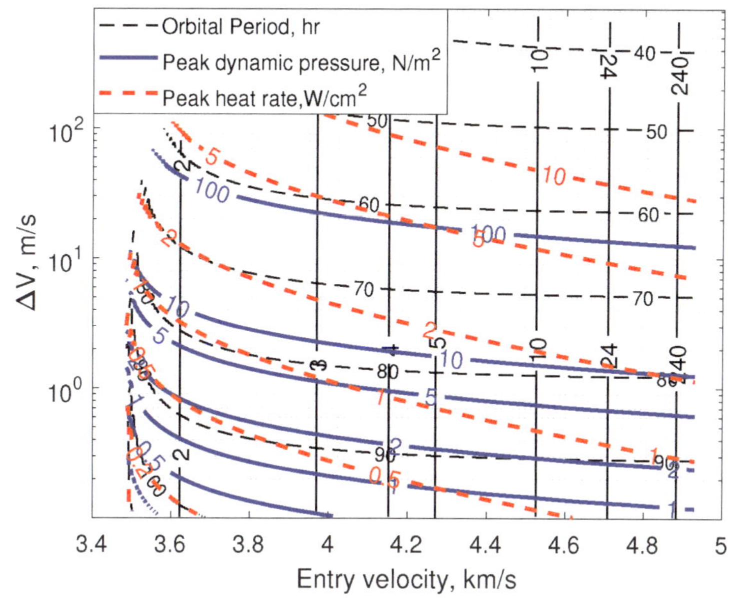

Figure 5.
Performance and design parameters for aerobraking maneuver for vehicle ballistic coefficient of 900 kg/m².

time for vehicles with a lower ballistic coefficient. Using the same timeline (i.e., 6 months) as the past robotic aerobraking missions, human-class missions would require a minimum dynamic pressure limit of 4.5 N/m^2. For faster aerobraking missions, a 1-month aerobraking maneuver requires a dynamic pressure limit of 25 N/m^2; a 10-day aerobraking maneuver requires 75 N/m^2. Note that at higher dynamic pressure, the heat rate constraint may be more dominating, and the numbers are only rough estimates for the purpose of illustrating the application of the plots.

4. Atmospheric entry at Mars

Mars atmospheric entry, descent, and landing are very challenging due to its thin atmosphere [1]. It is difficult for large entry vehicles to achieve enough deceleration for a soft touchdown. Landers and rovers have been successfully delivered to the surface of Mars, but the mass class has been increasing—from 11 kg for Mars Pathfinder, 185 kg for Mars Exploration Rovers, to 900 kg for Mars Curiosity and 1050 kg for Mars 2020. As interests in human Mars missions increase, landing large-size human-rated payload (on the order of 10s of metric ton) on Mars surface becomes important. The main objective of the entry phase is to decelerate the vehicle enough so that the vehicle can reach a low velocity at a high enough altitude for the final descent and landing phase to safely engage. In the following, we will show a parametric analysis for the performance of different vehicle designs, from ballistic vehicles (L/D = 0) to mid-L/D vehicles (L/D up t0 0.8) for human-class missions.

It is important to note that, for robotic missions, the landed mass has been in the range of 100s to 1000s kg, thus with a reasonably large aeroshell (e.g., diameter of 4.5 m), a ballistic coefficient of less than 50 kg/m^2 can be achieved. However, for human-class payloads, there is a need for mid-L/D vehicles, for which the ballistic coefficients are at least an order of magnitude larger than that of a robotic mission. To conduct the parametric analysis, we use 50 and 900 kg/m^2 as the baseline and show the results corresponding to these vehicle designs. Interested readers are directed to the Mars Design Reference Architecture 5.0 [22] for more details.

4.1 Entry performance parameters

Key performance parameters for atmospheric entry are peak g-load, peak heat rate, and total heat load. Lower values for all the parameters are desired; however, trends for peak heat rate and total heat load are opposite as shown in **Figure 6**. For the same entry velocity, with increase of entry flight-path angles, total heat load decreases whereas peak heat rate and peak g-load increases. As ballistic coefficient increases, an overall increase in all three parameters can be observed.

In **Figure 6**, the black shaded areas in all four plots correspond to exit trajectories, meaning that with the combination of velocity and entry flight-path angle, the vehicle will exit the atmosphere after a short atmospheric flight. Several entry velocities are also worth noting—3.49 km/s is the circular velocity at 129-km altitude, 4.93 km/s for V_∞ of 0 km/s, 7 km/s for V_∞ of 5 km/s, and 9 km/s for V_∞ of 7.5 km/s. A V_∞ of 7.5 km/s is the common maximum value seen for Mars arrival V_∞. Higher arrival velocities are possible but for the purpose of this chapter, an entry velocity of 9 km/s covers a wide range of interplanetary trajectories.

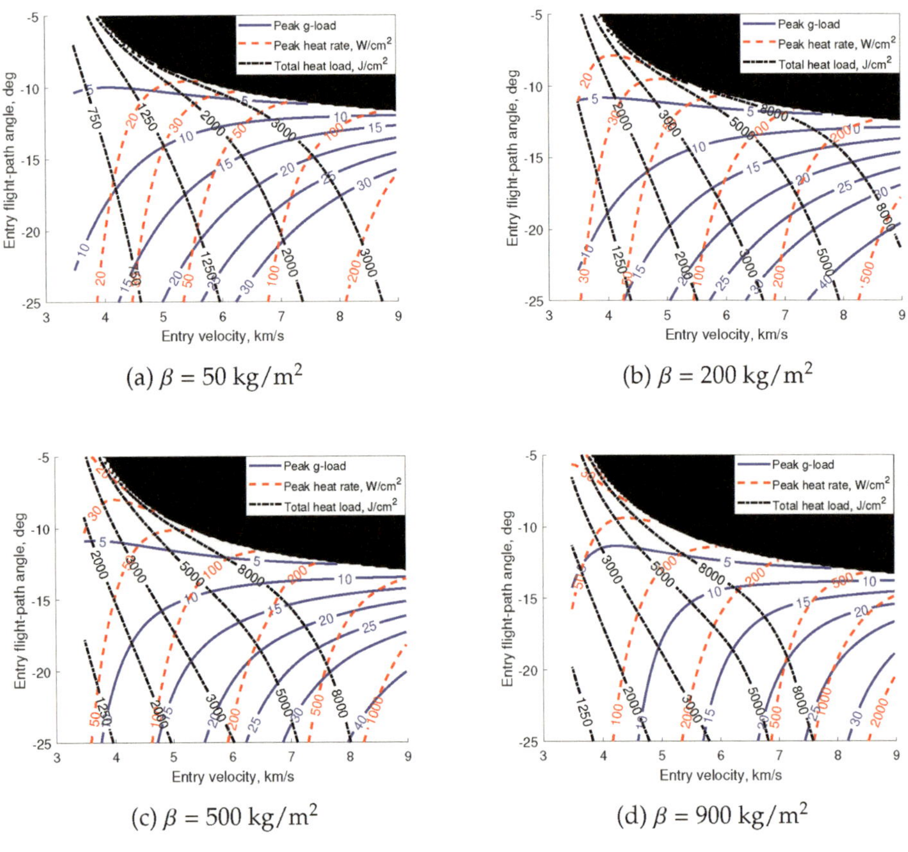

(a) $\beta = 50$ kg/m^2

(b) $\beta = 200$ kg/m^2

(c) $\beta = 500$ kg/m^2

(d) $\beta = 900$ kg/m^2

Figure 6.
Entry parameters for ballistic vehicle (L/D = 0), showing peak g-load, peak heat rate, and total heat load for entry velocity from circular speed at 129 km altitude to V_∞ of 7.5 km/s.

At entry flight-path angle of about −11 deg, **Figure 6** shows that the peak g-load only varies slightly with entry velocities, which means that entry at both high arrival velocities and low velocities results in similar g-load at that particular entry flight-path angle. However, the main differences will be the heat rate and heat load. If the mission design allows for a lower arrival velocity, it will decrease both peak heat rate and total heat load.

As a reference for technologies for entry, in particular TPS materials, we have listed some common TPS materials in **Table 3**. Note that heat rate on the order of 100 s W/cm^2 is well within the technology of current TPS material. Phenolic Impregnated Carbon Ablator (PICA) and Heatshield for Extreme Entry Environment Technology (HEEET) are more capable compared with other TPS materials. From **Figure 6**, for robotic missions, that is, ballistic coefficients of 50 and 200 kg/m^2, ACOAT could handle most of the heating conditions.

4.2 Terminal velocity

Terminal velocity at the end of entry phase during an EDL sequence is important for Mars. If the vehicle did not have enough deceleration from the initial entry phase, the descent and landing will not have enough time for execution, eventually leading to a crash. Successful Mars missions all use parachute as a means for descent. **Table 4** lists the altitudes and Mach numbers for the start of final descent phase (i.e., parachute deployment). Note that, each mission delivered the probe to a different altitude on Mars; while the requirements can be different for each mission, the numbers in **Table 4** provide a guideline for the desired mach numbers at the end of the entry phase.

Parachute is not the only means for descent, as high-mass class vehicles are emerging for human missions, the feasibility of a large enough parachute for descending 10s metric tons of payloads are questionable. For such reason,

Material	Density, g/cm^3	Maximum $\dot{q}_s$, W/cm^2	Pressure, atm
Shuttle tiles	0.192–0.352	44	
SRAM family	0.224–0.32	~100	~1
TUFROC	Varies	~300	—
SLA-561V	0.256	100–200	<0.3
AVCOAT		~900	~1
PICA	0.256	>1400	~0.3
HEEET	0.3–1.4	~7000 (tested [24])	—

Table 3.
Properties and performances of TPS materials [8, 9, 23].

Mission	Mach number	Altitude, km	Landing site elevation, km
1976 Viking 1 & 2	1.1	5.79	−3.5
1997 Pathfinder	1.57	9.4	−2.5
2004 MERs	1.77	7.4	−1.9
2012 Curiosity	2.05	10.0	−1.45

Table 4.
Parachute deployment altitude and Mach number.

supersonic retro-propulsion is an attractive option for high-mass systems [25], which could be activated at Mach 2 and above.

To evaluate the overall performance of the terminal velocity of entry systems, **Figure 7** shows the terminal velocities (at 10 km) for small (50 kg/m^2) and large (900 kg/m^2) ballistic coefficients, corresponding to robotic and human missions respectively. Shallow entry flight-path angles are preferred in order to achieve a lower terminal velocity to ensure a safe descent phase. For ballistic coefficient of 50 kg/m^2, the lowest terminal velocity is about 1.7 km/s. The contour line denoting "1.8" in **Figure 7a** shows that the same terminal velocity can be achieved at two different entry flight-path angles. Such phenomena are due to a partial equilibrium glide phase, which is shown in **Figure 8**. The difference between the two trajectories is the downrange, which can be problematic, because any uncertainties in entry condition, atmospheric densities, or aerodynamic properties will cause the vehicle to deviate from the planned trajectory.

For practical design consideration, if pinpoint landing is required, lifting body vehicle is a must in order to actively control the trajectory to the target areas. Mars Science Laboratory is an example where it has a targeted landing site for scientific

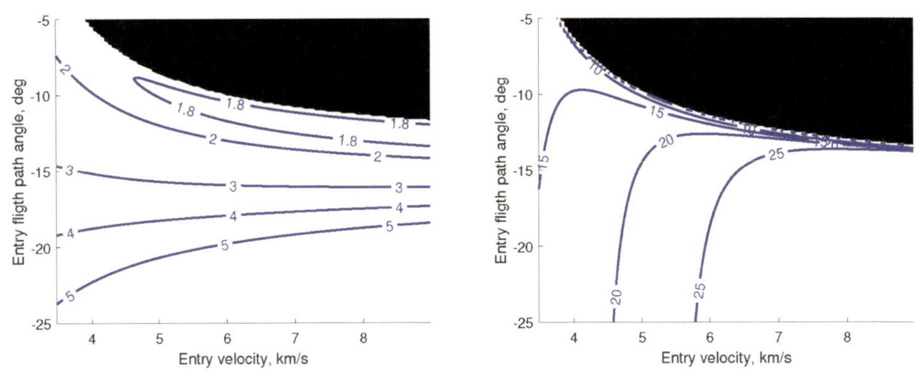

(a) $\beta = 50$ kg/m^2 (b) $\beta = 900$ kg/m^2

Figure 7.
Terminal velocities in Mach number for ballistic vehicle (L/D = 0) at an altitude of 10 km.

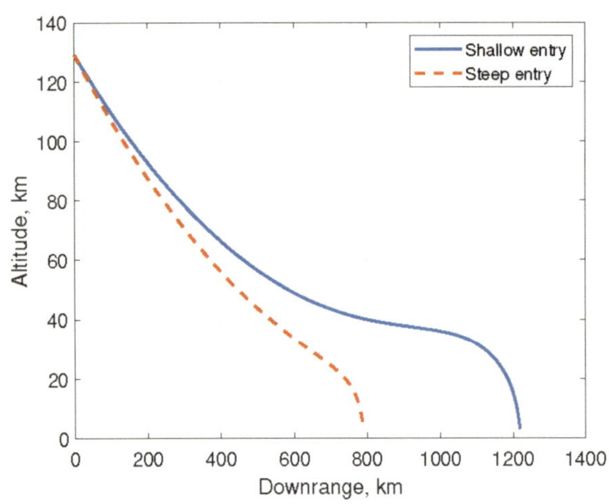

Figure 8.
Ballistic entry trajectories with the same terminal velocity for shallow and steep entry.

explorations. The same landing requirement will also apply for human missions due to the requirement of landing close to the base locations. A guided entry becomes ever important so that at the end of entry phase the vehicle can be within the targeted landing site and with the delivery range of the final descent stage. **Figure 9** shows that adding lifting capability can significantly reduce the terminal velocity compared with **Figure 7a**. For any targeted entry flight-path angles, the terminal velocity at 10 km ranges from Mach 1.3 to 1.5. A vehicle with higher L/D could achieve even lower terminal velocity. The catch however is that, most of these trajectories would leverage the lifting capability to fly higher in altitude to achieve more deceleration.

Shown in **Figure 10** are the trajectory profiles for the same terminal velocity at different entry flight-path angles. A portion of the trajectory skip up thus results in more deceleration. Note that **Figures 9** and **10** are for fully lift-up trajectories only, no guidance has been implemented. The results show that adding lifting capability

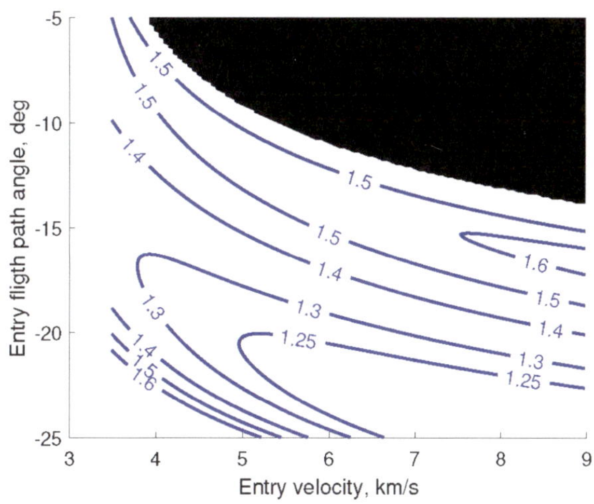

Figure 9.
Terminal velocities in Mach number for vehicle L/D of 0.2 at an altitude of 10 km, β = 50 kg/m^2.

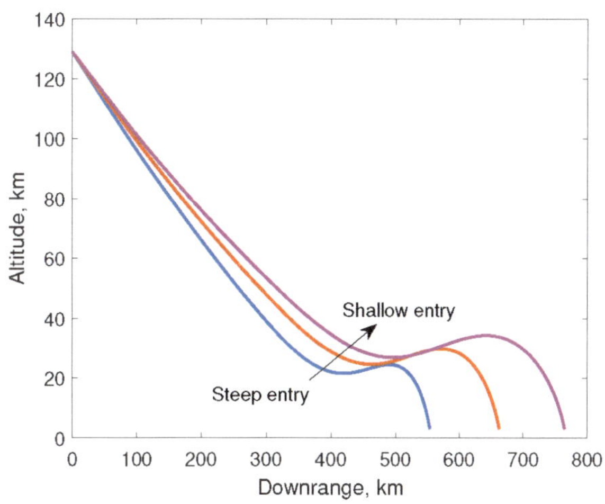

Figure 10.
Trajectory profiles for the same terminal velocity at different entry flight-path angles, using vehicle L/D of 0.2.

can significantly reduce the terminal velocities. Vehicles with higher L/D can achieve even lower terminal velocity due to increased control authority.

5. Aerocapture at Mars

Mars aerocapture has been popular for both small satellite missions and human missions. For small satellite missions, due to the limited mass budget, propulsion system has propellant mass requirement that is too restrictive, and traditional bank modulation requires very complicated thrusters for banking maneuver. Drag modoluation aerocapture has been considered for Mars small satellite missions [26]. Methods for comparing aerocapture with traditional propulsive options and aerobraking were also investigated for Venus in Ref. [27]. For human Mars missions, repeated Mars orbit insertion maneuvers warrant the use of aerocapture to save propellant in order to deliver assets to Mars orbit.

Since aerocapture is an orbit insertion maneuver, the atmospheric entry velocities correspond to V_∞ of 0–7.5 km/s, which is similar to the velocity ranges used for entry, but with the exclusion of entry velocities for closed orbits (which is also the range of entry velocities used for aerobraking analysis).

Aerocapture maneuver is very sensitive to uncertainties; thus in order to ensure a successful aerocapture maneuver, vehicle control authority is required. A measure of such control authority is termed theoretical corridor width, which is measured in deg. Theoretical corridor width is the difference between the maximum and minimum entry flight-path angles that allow the vehicle to be captured given a set of parameters, including arrival V_∞, vehicle L/D, and ballistic coefficients. Another term of corridor width is the required corridor width, which is a measure of the total uncertainties in deg. The uncertainties mainly include atmospheric density uncertainties, vehicle aerodynamic uncertainties, and arrival uncertainties (i.e., target B-plane and entry flight-path angle).

5.1 Aerocapture feasibility

A framework for assessing aerocapture feasibility has been developed and discussed in detail in Ref. [10, 27, 28] and the results in this section follow the same analysis framework. The key design parameters for aerocapture missions are very similar to those for entry except theoretical corridor width. Peak g-load, peak heat rate, and total heat load are the parameters of interests for aerocapture maneuver. **Figures 11** and **12** combine all the design parameters as contours lines onto a single plot including arrival V_∞ and vehicle L/D.

It is important to recognize the implications of the contours. The solid contour "Corridor width" notes the theoretical corridor width and it has a lower limit, which is the required corridor width. For Mars, the required corridor width is conservatively estimated to be 2 deg. [29]. Given that theoretical corridor width has to be greater than the required corridor width, the area under the solid contour of 2 deg. is the unfeasible region for aerocapture due to the lack of control authority. Then the design parameters—peak g-load, peak heat rate, and total heat load—place upper constraints on the aerocapture feasibility. For example, if the mission requires a maximum g-load of 10 Earth g's for a human mission, then we can tell from the feasibility plot, **Figure 11**, that there exists a maximum arrival V_∞ and that the mission designer should only search for trajectories that result in a lower arrival V_∞ than the limit.

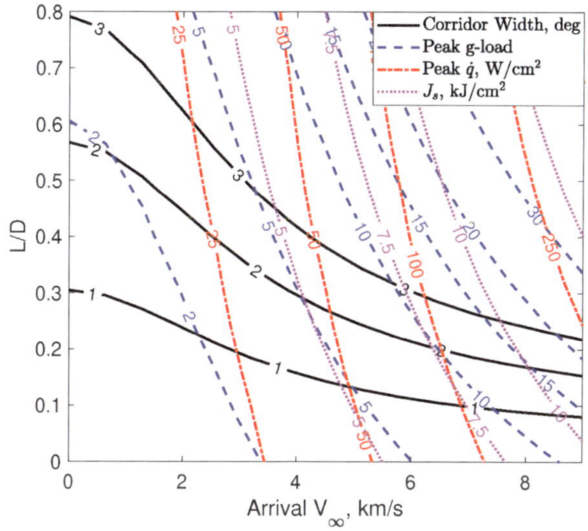

Figure 11.
Aerocapture feasibility plot for $\beta = 50$ *kg/m²*.

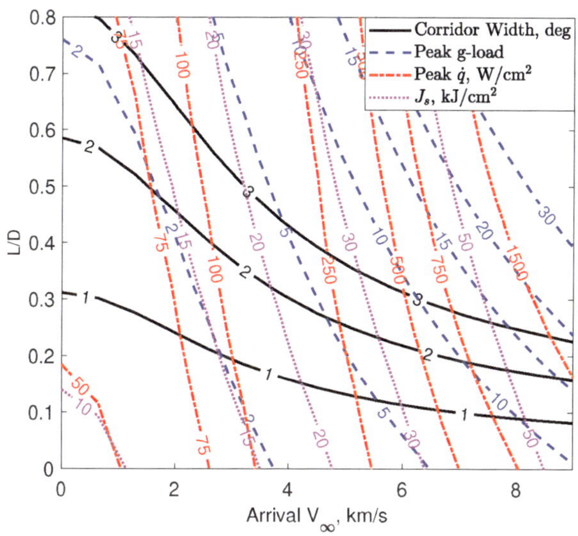

Figure 12.
Aerocapture feasibility plot for $\beta = 900$ *kg/m²*.

A common trend of all contours of the design parameters is that they increase with the increase of arrival V_∞ and with increase of vehicle L/D. It is intuitive that higher velocity will result in higher g-load, heat rate, and total heat load. However, the trend for vehicle L/D is due to the assumption used in the analysis that only the worst-case scenarios are used. For a value of vehicle L/D, both full lift-up and full lift-down cases are evaluated and the worse case of the two was recorded and plotted. The design values shown are the worst-case scenarios and it is expected that with guidance and control, the actual values will be lower.

The contours of peak g-load, peak heat rate, and total heat load put constraints on the plots. Similar to how g-load constrains the design feasibility, both peak heat rate and total heat load will constrain the feasible arrival V_∞. As the contour lines of these design constraints intersect with the theoretical corridor width, there also exists a minimum vehicle L/D for a successful aerocapture. As the constraints become more restrictive (in other words, allowable peak g-load and peak heat rate are reduced), the requirement for vehicle L/D will increase whereas the maximum allowable arrival V_∞ will decrease.

5.2 Aerocapture for robotic and human missions

The difference between **Figures 11** and **12** is the ballistic coefficients. By comparing the two plots, we note that a higher ballistic coefficient results in similar theoretical corridor width (i.e., vehicle control authority), similar peak g-load, but much higher peak heat rate and total heat load. **Figure 11** can be regarded for robotic missions or small satellite missions whereas **Figure 12** for human Mars missions.

Figure 11 shows that the peak heat rate and total heat load are very benign. Even at V_∞ of 9 km/s, the peak heat rates are only around 250 W/cm^2, which is well within the TPS material limits. At lower V_∞, we can even use non-ablative TPS materials such as the tiles used on Space Shuttles as listed in **Table 3**. In terms of peak g-load, robotic missions can usually tolerate a higher g-load than for human missions. The Galileo probe that entered Jupiter's atmosphere was designed to withstand over 200 Earth g's. As a result, in the range of arrival V_∞ considered, the g-load constraints are not too restrictive.

For human Mars missions, as shown in **Figure 12**, the peak heat rates are still well within the current TPS materials. As with higher V_∞ values, peak heat rate can be more challenging. However at around 2000 W/cm^2, HEEET is capable of handling such heat rate. Note that HEEET is an ablative TPS material, which means that it will be difficult to reuse because of the loss of materials. For the purpose of getting humans to Mars, a non-ablative TPS material will be ideal for repeated aerocapture maneuvers. Considering a theoretical corridor width of 2 deg., with a mid-L/D vehicle (L/D of 0.6–0.8), the peak heat rate will be around 75 W/cm^2, which is more than what the shuttle tile can handle. Assuming a non-ablative TPS material can sustain 75 W/cm^2 peak heat rate; then, in **Figure 12**, the area above the contour line of 2 deg. and left of 75 W/cm^2 line is the feasible region for aerocapture, which requires a very low arrival V_∞ of less than 1.5 km/s. In terms of the g-load constraints for human missions, within the areas found, a peak g-load of 2 Earth g's is a very benign condition. It is important to note again, that the worst-case scenarios are shown, and with guidance and control, optimal trajectories can often reduce the peak conditions.

6. Conclusions

In this chapter, a high-level assessment of aerocapture, aerobraking, and entry for robotic and human mission is presented. A comprehensive parametric analysis for all three maneuvers has been investigated while considering the key design parameters. Vehicle aerodynamic properties are key drivers in the performance of these maneuvers. Entry velocities also affect greatly the design parameters. From the results, aerocapture, aerobraking, and entry can be successful for robotic missions, whereas for human Mars missions, there are still challenges that need to be addressed. The challenges are directly related to the risks of the mission, and for human missions, safety is usually the top priority and strategies to mitigate the risk, that is, addressing the challenges will significantly reduce the risk to ensure mission success.

Author details

Ye Lu
Kent State University, Kent, OH, United States

*Address all correspondence to: ylu16@kent.edu

References

[1] Braun RD, Manning RM. Mars exploration entry, descent and landing challenges. Journal of Spacecraft and Rockets. 2007;**44**(2):310-323

[2] London HS. Change of satellite orbit plane by aerodynamic maneuvering. Journal of the Aerospace Sciences. 1962; **29**(3):323-332

[3] Spencer DA, Tolson R. Aerobraking cost and risk decisions. Journal of Spacecraft and Rockets. 2007;**44**(6): 1285-1293

[4] Vinh NX, Busemann A, Culp RD. Hypersonic and Planetary Entry Flight Mechanics. Ann Arbor, MI, USA: The University of Michigan Press; 1980

[5] Sutton K, Graves RA. A general stagnation-point convective-heating equation for arbitrary gas mixtures. In: Technical Report. Hampton, VA, USA: NASA; 1971

[6] Tauber ME, Suttont K, Sutton K. Stagnation-point radiative heating relations for Earth and Mars entries. Journal of Spacecraft and Rockets. 1990; **28**(1):40-42

[7] Knittel JM, Lewis MJ. Concurrent arocapture with orbital plane change using starbody waveriders. The Journal of the Astronautical Sciences. Springer; 2014;**61**(4):319-340

[8] Lockwood MK, Edquist KT, Starr BR, Hollis BR, Hrinda GA, Bailey RW, et al. Aerocapture systems analysis for a Neptune Mission. In: Technical Report. NASA/TM-2006-214300. Hampton, VA, USA: NASA; 2006

[9] Lockwood MK, Queen EM, Way DW, Powell RW, Edquist K, Starr BW, et al. Aerocapture systems analysis for a Titan mission. In: Technical Report NASA/TM-2006-214273. 2006

[10] Lu Y, Saikia SJ. Feasibility assessment of aerocapture for future titan orbiter missions. Journal of Spacecraft and Rockets. 2018;**55**(5): 1125-1135

[11] Venkatapathy E, Arnold J, Fernandez I, Hamm KR, Kinney D, Laub B, et al. Adaptive deployable entry and placement technology (ADEPT): A feasibility study for human missions to Mars. In: 21st AIAA Aerodynamic Decelerator Systems Technology Conference and Seminar, (Dublin, Ireland), AIAA. 2011

[12] Hughes S, Cheatwood F, Dillman R, Calomino A, Wright H, DelCorso J, et al. Hypersonic inflatable aerodynamic decelerator (HIAD) technology development overview. In: 21st AIAA Aerodynamic Decelerator Systems Technology Conference and Seminar, (Dublin, Ireland), AIAA. 2011

[13] Henning GA, Edelman PJ, Longuski JM. Design and optimization of interplanetary aerogravity-assist tours. Journal of Spacecraft and Rockets. 2014;**51**(6):1849-1856

[14] Putnam ZR, Braun RD. Drag-modulation flight-control system options for planetary aerocapture. Journal of Spacecraft and Rockets. 2014; **51**(1):139-150

[15] Putnam ZR, Braun RD. Precision landing at Mars using discrete-event drag modulation. Journal of Spacecraft and Rockets. 2014;**51, 1**:128-138

[16] Lafleur J, Cerimele C. Angle of attack modulation for Mars entry terminal state optimization. In: AIAA Atmospheric Flight Mechanics Conference, Vol. 8, August, Reston, Virigina. American Institute of Aeronautics and Astronautics. 2009. pp. 1-20

[17] Queen EM. Angle-of-attack-modulated terminal point control for Neptune aerocapture. In: 14 AAS/AIAA Space Flight Mechanics Conference, Maui, Hi. AAS. 2004. pp. 04-129

[18] Dwyer Cianciolo A, Powell RW. Entry, descent, and landing guidance and control approaches to satisfy Mars human mission landing criteria. In: AAS/AIAA Flight Mechanics Conference. 2017

[19] Fujino T, Yoshino T, Ishikawa M. Numerical analysis of reentry trajectory coupled with magnetohydrodynamics flow control. Journal of Spacecraft and Rockets. 2008;**45**:911-920

[20] Shibata E. Aeropropcapture: Applying propulsion to aerocapture maneuvers. In: AIAA/AAS Astrodynamics Specialist Conference. 2018

[21] Giersch L, Knarr K, Lyons DT. The dedicated deployable aerobraking structure. In: Technical Report. Pasadena, CA, USA: Jet Propulsion Laboratory; 2011

[22] Drake GG, Yoder GL, McCuistion D, Irvine TB, Radzanowski DP. Human exploration of Mars design reference architecture 5.0. In: Technical Report. Hampton, VA, USA: NASA; 2009

[23] Kolawa E, Balint T, Birur G, Bolotin G, Brandon E, Castillo LD, et al. Extreme environments technologies for future space science missions. In: Technical Report. Pasadena, CA, USA: NASA; 2007

[24] Ellerby D, Gage P, Kazemba C, Mahzari M, Nishioka O, Peterson K, et al. Heatshield for extreme entry environment technology (HEEET) development status. In: 13th International Planetary Probe Workshop, Laurel, MD. 2016

[25] Edquist KT, Desai PN, Schoenenberger M. Aerodynamics for Mars Phoenix entry capsule. Journal of Spacecraft and Rockets. 2011;**48**:713-726

[26] Putnam ZR, Clark IG, Braun RD. Drag modulation flight control for aerocapture. In: IEEE Aerospace Conference. 2012

[27] Girija AP, Lu Y, Saikia SJ. Feasibility and mass-benefit analysis of aerocapture for missions to Venus. Journal of Spacecraft and Rockets. 2020; **57**:58-73

[28] Saikia S, Millane J, Mudek A, Arora A, Witsberger P, Shibata E, et al. Aerocapture assessment at Uranus and Neptune for NASA's ice giant studies. In: Technical Report, PU-AAC-2016-MC-0002. West Lafayette, IN: Purdue University; 2016

[29] Wright HS, Oh DY, Westhelle CH, Fisher JL, Edquist KT, Brown JL, et al. Mars aerocapture systems study. In: Technical Report, NASA/TM-2006-214522 Mars. 2006

Chapter 4

Assessing Propulsion and Transportation Issues with Mars' Moons

Bryan Palaszewski

Abstract

This chapter is focused on transportation issues with Mars' moons: Phobos and Deimos. The moons are small nonspherical bodies that may offer unique specimens for science, a gateway to understanding the asteroid belt, and resource platforms for space industries. The mission delta-V and both chemical propulsion and nuclear electric propulsion (NEP) orbital transfer vehicles (OTVs) are analyzed. The use of nuclear electric propulsion allows very large reductions on the resupply propellant mass over chemical propulsion options. Large delta-V plane changes are also more efficient using electric propulsion. The benefits of electric propulsion are unique, and the power system can support high-power radar science experiments.

Keywords: in situ resource utilization, ISRU, moon base, rocket propulsion, systems analysis, specific impulse, chemical propulsion, nuclear propulsion, electric propulsion

1. Introduction

Mars is the fourth planet from the Sun. Its environment includes a 95% carbon dioxide atmosphere with a very low pressure (5–7 millibar), and essentially no magnetosphere, though there are remnants of the magnetic fields in small area of the planet. Exploration programs for Mars have included robotic and human surface visits and human bases. Mars has two moons: Phobos and Deimos. The moons are small and akin to asteroids. They can be a great source of materials for exploration and exploitation.

The Martian moons are tantalizing objects for scientific investigation. The moons also present a unique set of challenges. Their surface gravity is very low: 8.7×10^{-4} for Phobos and 6.2×10^{-4} Earth gravities (g) for Deimos, as shown in **Table 1** [1–3]. The gravity levels are computed based on an average diameter, as the moons are nonspherical. Based on their shape and features, both moons may be captured asteroids. Science measurements of the moon's structure may lead to a better understanding of the diversity of asteroids in our solar system. Based on spectral analyses, both moons may contain carbonaceous chondrites, other metals, and water. Such materials can be the resources to propel fledgling in situ resource utilization (ISRU) industries.

Phobos has a giant crater, Stickney, which is 9 km in a large fraction of the moon's diameter. Deep grooves cover the tiny moon [3]. The crater dynamics have fascinated geologists and planetary impact modelers alike. Photos of the moons are shown in Appendix A.

Body	mu (Km^3/s^2)	R (km)	G	m (Kg)	a (m/s^2)	g Level
Mars	4.28E+04	3396.20	6.67E−17	6.42E+23	3.71E+00	3.79E−01
Phobos	7.07E−04	9.1	6.67E−17	1.06E+16	8.54E−04	8.72E+04
Deimos	1.60E−04	5.1	6.67E−17	2.40E−15	6.16E−03	6.29E−04

Table 1.
Gravity levels of Mars and its moons.

Any scientific investigations will likely include radar studies to determine the moons' internal structures and surface sampling. Small robotic landers will likely be precursors to human landings. As the moon does not have a uniform shape, the gravity level on different surface locations will vary.

The Martian moon overall characteristics and surface gravity are presented in **Table 1**. The gravity level is computed using the smallest dimension of each moon. Previous missions have sought to rendezvous, orbit, or fly by the moons. As with some comets (67P; [4, 5]), the orbital mechanics may necessitate a propulsive station keeping above the moons. The moons' low gravity levels have led research on anchoring technologies for landing vehicles [6–11].

While the moons potentially have water resources, the complexity of extracting the water may be daunting. The low moon gravity will necessitate the use of unique capturing technologies. The low gravity will allow regolith to be liberated and potentially create a dangerous or at least a complicated dust environment. Large boulders may be a more controllable source for regolith processing. The gravity levels of outer planet moon (Naiad) of Neptune are similar to those of Phobos and Deimos. Outer planet moon analyses [12] have suggested using an artificial gravity space base for high value ISRU material processing. Such a factory might reside near the moon or be anchored to its surface. The regolith might be fed into the factory, and the artificial gravity system with the appropriate thermal energy would assist in separating the water resources from the dust and rock. Investigating several mining methods for extremely low gravity moons will be essential for any successful ISRU architecture.

2. Mission design and options

Phobos and Deimos exploration methods have been studied for many decades: landers, flybys, etc. [6–11]. While landers have been assessed in the past, this chapter will focus on the orbital transfer delta-V requirements and orbital transfer vehicle (OTV) designs that would allow the two moons' exploration and exploitation.

Three general missions were assessed: flights from the moon's orbit to Mars orbit (LMO), flights between the two moons, and flights from the moon's orbit to a very high Mars orbit (100,000 km altitude). A fourth mission delta-V, for transfer to the areosynchronous Mars orbit, was also computed (**Table 2**).

Additional delta-V calculations for missions to high inclination orbits were also investigated. The high inclinations may be attractive for polar monitoring or specialized payloads for surface observations, atmospheric studies, and interplanetary communications or power satellites.

Both high-thrust missions and low-thrust missions were assessed. The high-thrust delta-V values were computed with a standard Hohmann transfer equations [13]. The values for the low-thrust delta-V were calculated using the Edelbaum equation (Ref. [14]). The nominal semi-major axes for Phobos and Deimos are 9378 and 22,459 km [2].

Figures 1 and **2** depict the round-trip delta-V for Phobos and Deimos missions, respectively. Both high-thrust and low-thrust delta-V values are presented. Due to the typical gravity losses with high-thrust propulsion systems, a 20% delta-V increase is added; no added losses were imposed on the low-thrust systems. In **Figure 1**, the highest value is the Phobos to 100,000 km delta-V 2.99 km/s. The Phobos to low Mars orbit (LMO) delta-V was 2.74 km/s. The LMO altitude is 100 km. At Deimos, the highest round-trip delta-V is for the Deimos to LMO transfer, 4.3 km/s. The transfer to 100,000 km requires only 1.42 km/s.

Figure 3 shows the high-thrust plane change delta-V values to reach high inclinations while performing the plane changes at the orbital altitudes of Phobos and Deimos. In **Figure 4**, delta-V for the inclination change being performed, the high altitude of 100,000 km is presented. The two-way delta-V for the Phobos or Deimos orbital transfer to 100,000 km would have to be added to the values in **Figure 4**. These two cases are presented, in that if the very high inclination changes are performed at high altitude, the total mission's delta-V is reduced over the low-altitude inclination changes.

Payload flights—Phobos and Deimos
Access low Mars orbit (LMO)
Access all orbital inclinations
Access areosynchronous Mars orbit (AMO)
Deliver and recover high altitude payloads
Resupply factories
Carry ISRU propellants
Carry ISRU products, other than propellants

Table 2.
Mars moon's orbit payload options and delivery destination.

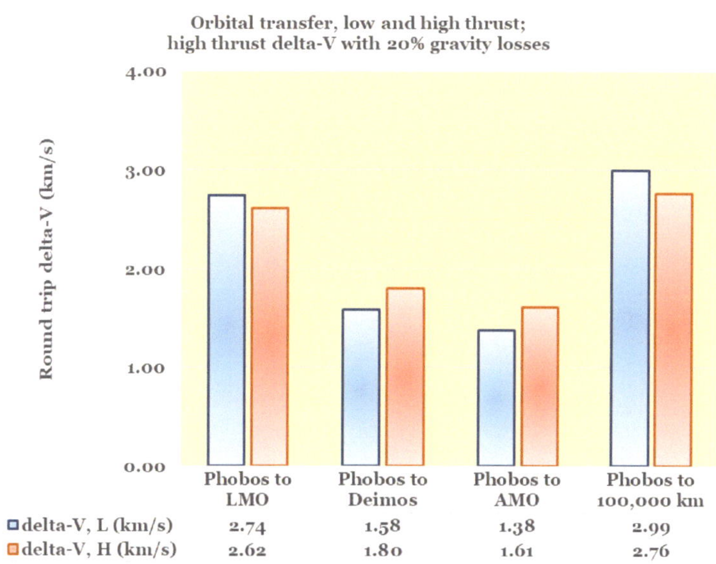

Figure 1.
Mission options and delta-V—Phobos, using low thrust (blue) and high thrust (orange).

3. Propulsion options

High-thrust chemical propulsion, using oxygen/hydrogen (O_2/H_2) rocket engines is a natural choice [12]. If indeed water were available on the Martian moons, it would make sense to capitalize on that water resource.

Electric propulsion systems with either ion or Hall thrusters are potential options. Xenon or other inert gases are the typical choice for such thrusters. Using hydrogen as an electric propulsion propellant has also been proposed [12]. However, the hydrogen propellant option is a far term prospect [12].

Mass scaling equations were developed for the O_2/H_2 and the nuclear electric propulsion (NEP) systems [12].

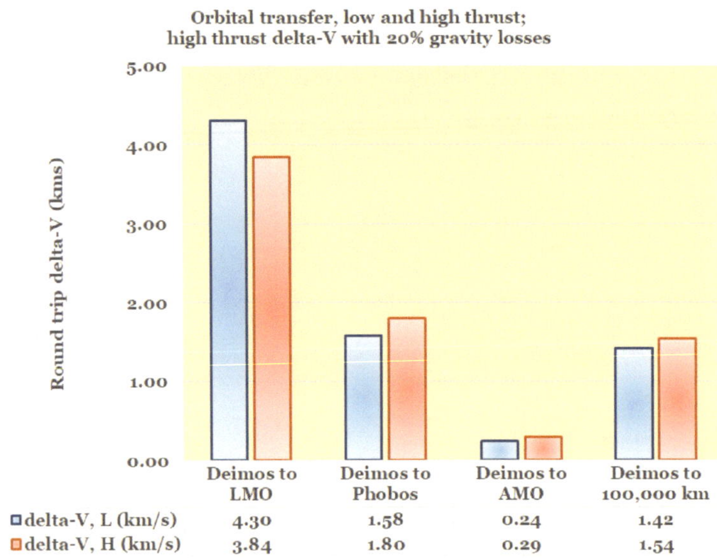

Figure 2.
Mission options and delta-V—Deimos, using low thrust (blue) and high thrust (orange).

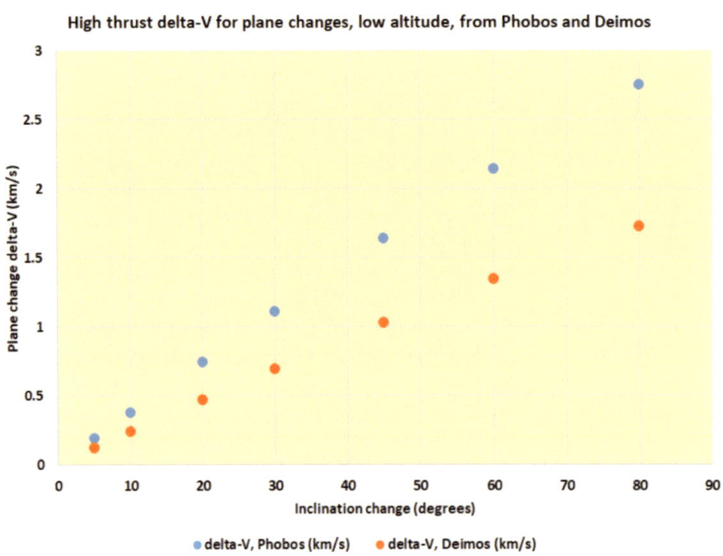

Figure 3.
Orbital transfer at low altitude—Phobos and Deimos altitude.

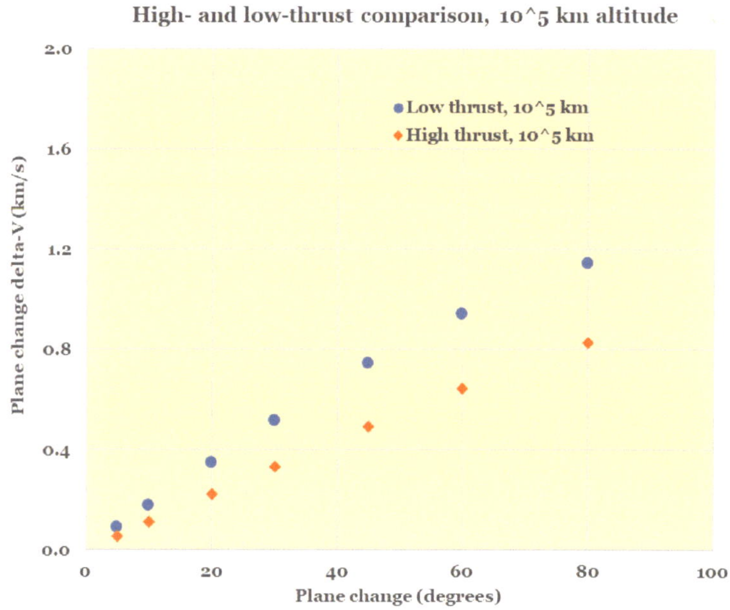

Figure 4.
Orbital transfer—high-altitude plane change, 100,000 km.

3.1 Advanced propulsion options

Several advanced propulsion options for lunar base construction and industrialization were investigated. They include nuclear electric propulsion options, lunar base design options, propellant industrialization, and outer planet mining with associated outer planet moon bases. Chemical propulsion and nuclear electric propulsion (NEP) for Earth-Moon orbital transfer vehicles (OTVs) were assessed. Design parameters, vehicle mass scaling equations, and summaries of these analyses are presented.

3.1.1 Chemical propulsion OTV sizing

In sizing the chemical propulsion OTVs, a vehicle mass scaling equation is used [12]:

$$m, dry, stage = m, dry, coefficient * m, p + a, fixed \tag{1}$$

where
m, dry, stage = the stage dry mass, including residual propellant (kg);
m, dry, coefficient = the B mass coefficient (kg of tank mass/kg of usable propellant mass);
m, p = usable propellant mass (kg); and
a, fixed = chemical OTV fixed mass (kg).

The chemical propulsion OTVs had a B coefficient of 0.4. The fixed mass was 500 kg. The fixed mass includes guidance systems, adapters, and reaction control system masses. The Martian moon OTVs were single-stage vehicles.

3.1.2 NEP OTV sizing

The NEP OTV mass and trip time were estimated based on the power system and the propulsion system design [11]. The following dry mass scaling equation was used [11]:

$$m,dry,stage,NEP = alpha * P + 0.05m,p + m,fixed \quad (2)$$

where
m, dry, stage, NEP = NEP dry mass (kg);
alpha = NEP reactor specific mass (kg/kWe);
P = NEP power level (kWe);
0.05 = tankage mass coefficient (kg/kg m, p);
m, p = NEP usable propellant mass (kg); and
m, fixed = NEP fixed mass (kg).

The OTV sizing was conducted for a wide range of power levels: 0.5–30 MWe. Three nuclear reactor-specific masses were used: 10, 20, and 40 kg/kWe (kilograms per kilowatt, electric). The OTV propulsion fixed mass, apart from and in addition to the reactor mass, was 20 MT, and the propellant tankage mass was 5% of the mass of the required propellant.

The Isp and efficiency of the electric propulsion systems were 5000 seconds with overall thruster propulsion efficiencies of 50% for each design. These design points are typical of advanced designs of either magnetoplasmadynamic (MPD) or pulse inductive thrusters (PIT). While hydrogen is suggested for both propulsion system thrusters, the possibilities of the higher Isp option using inert gases (xenon, krypton, etc.) are also viable. The low-thrust OTV delta-V value varied based on the destination of the Martian moon missions.

4. Mission effectiveness

4.1 Phobos and Deimos payload missions

Figures 5 and **6** depict the Phobos and Deimos O_2/H_2 propulsion system initial masses. For the 50 MT payload case for Phobos, the OTV initial mass is 141 MT. Nearly the same OTV mass is needed to perform the Phobos to LMO and Phobos to 100,000 km. For Deimos, the highest OTV mass is 256 MT.

The payload mass cases presented range from 1, 10, 20 to 50 MT. If the payload mass is less than 10 MT, the O_2/H_2 OTV mass is very small. If small 1 MT payloads must be sent quickly from one orbit to another, the O_2/H_2 OTV is an excellent choice; the propellant mass of the chemical propulsion system is very low compared to the NEP propellant mass. Alternatively, if five 10 MT payloads can be manifested together, the NEP OTV has a significant propellant mass advantage over the O_2/H_2 OTV.

Figures 7–9 present the round-trip mission trip times for the NEP vehicles for 1, 10, and 50 MT, respectively. The NEP trip times are many days long: for a 5 MWe NEP OTV with a 50 MT payload, the trip time for Phobos to LMO is 55 days, whereas the chemical propulsion trip times are less than 1 day. However, the benefits of reduced NEP propellant resupply mass are quite significant.

With NEP OTVs, the 10 MWe power levels provide the shortest trip time; however, if the payload can be delivered more slowly, the 1 MWe power level allows a very large propellant mass savings over the higher 10 MWe power level. The NEP propellant mass savings for large payloads are a critical part of any sustainable architecture. The propellant mass savings are noted in the succeeding sections.

Fast transfers of critical items under 1 MT are best accomplished with O_2/H_2 OTV propulsion. There may be a critical need for the delivery of medical supplies;

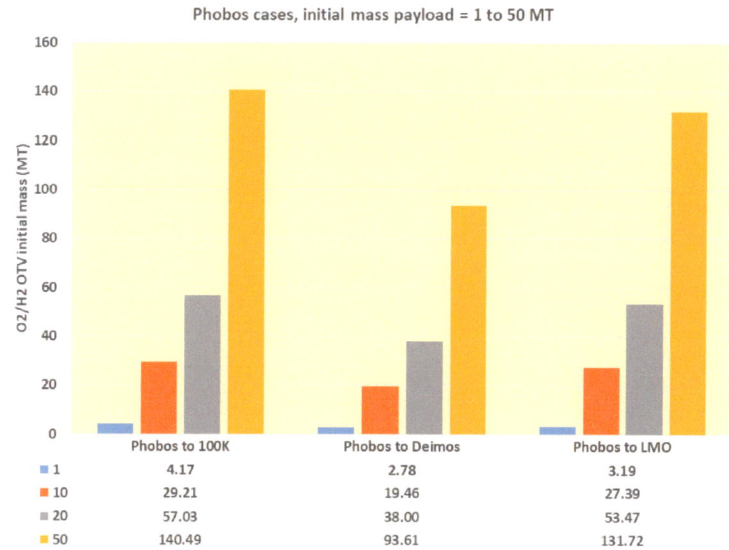

Figure 5.
Phobos OTV initial masses—O_2/H_2 propulsion.

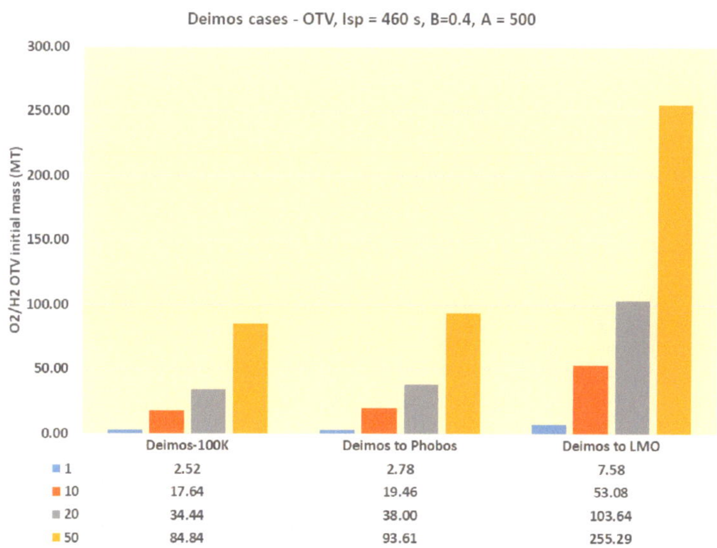

Figure 6.
Deimos OTV initial masses—O_2/H_2 propulsion.

also the delivery of space parts or a repair crew may be needed. The O_2/H_2 OTV would be best suited for these small 1 MT payloads.

Figures 10–13 compared the propellant masses for the O_2/H_2 system with several NEP systems. For the NEP cases, power levels of 0.5–10 MWe are shown. **Figures 10** and **11** present the Phobos and Deimos cases for 10 MT payloads, and the 50 MT payload cases are shown in **Figures 12** and **13**. In **Figure 10**, for a 10 MT payload, the Phobos to LMO NEP cases will allow large propellant savings over the O_2/H_2 OTV for NEP OTV power levels of less than 5 MWe. In the Deimos case, shown in **Figure 11**, the NEP OTV provides significant propellant mass savings over O_2/H_2 with power levels up to 10 MWe. The **Figure 12** data for Phobos to LMO with

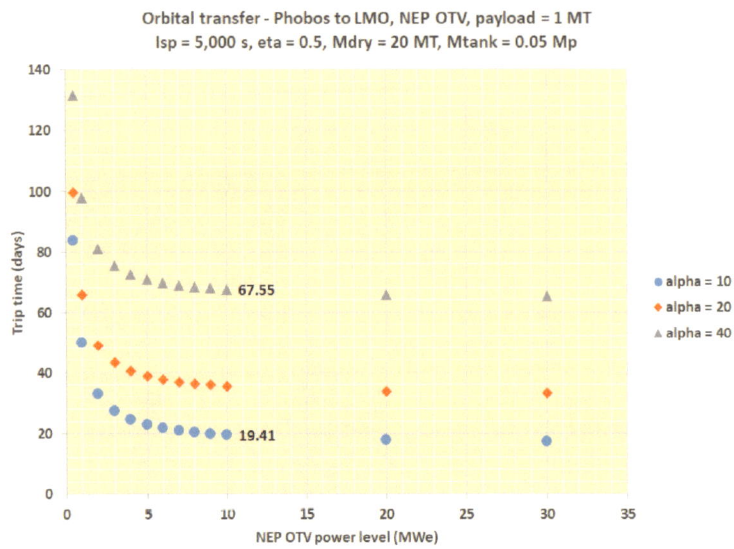

Figure 7.
NEP OTV trip time, 1 MT payload: Phobos to LMO.

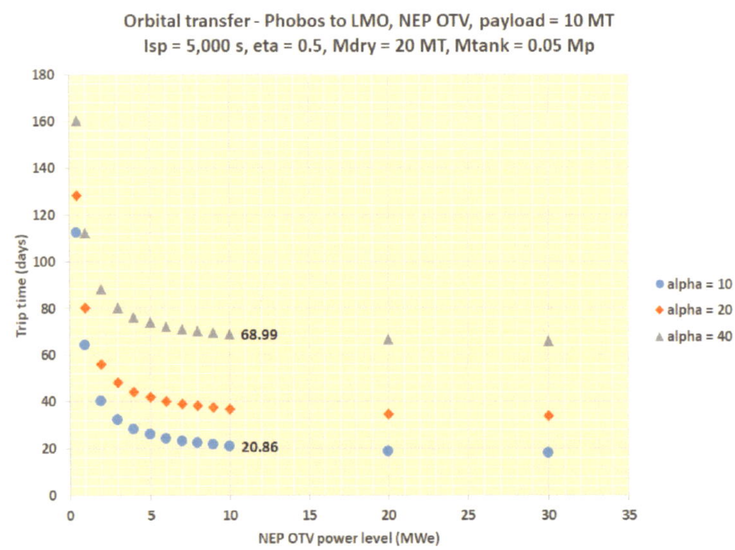

Figure 8.
NEP OTV trip time, 10 MT payload: Phobos to LMO.

a 50 MT payload shows very large NEP propellant mass reductions over O_2/H_2 for the 10 MWe power level; for a 5 MWe power level, the propellant mass reduction was from 57 to 10 MT. Similarly, in **Figure 13**, the Deimos to LMO cases show very significant propellant mass benefits, reducing the propellant needed by a factor of 6–10 or more over O_2/H_2. In nearly all cases, the NEP systems allow large propellant mass reductions. For large mission architectures over a long-term Mars project, the mass reductions can be as high as a factor of 5–10 over O_2/H_2 systems.

4.2 Mars lander options

Past studies of Mars landers have included an innovative single stage to orbit (SSTO) design [15]. The Mars Base Camp mission suggested an aerospacecraft that

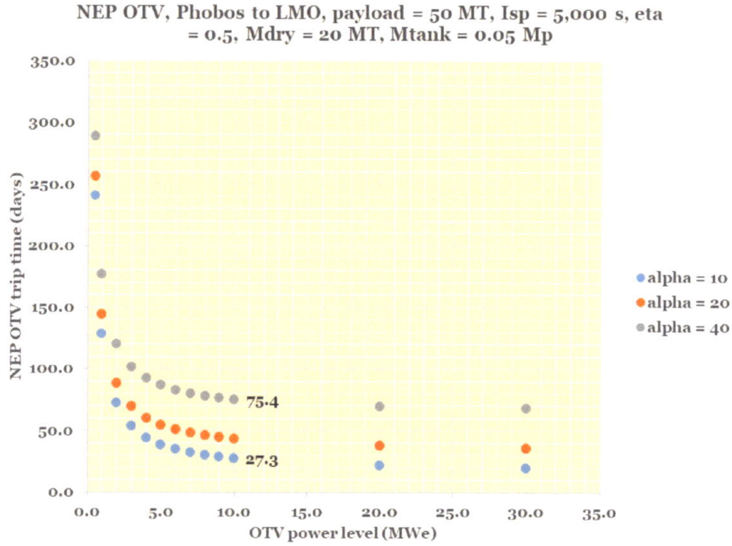

Figure 9.
NEP OTV trip time, 50 MT payload: Phobos to LMO.

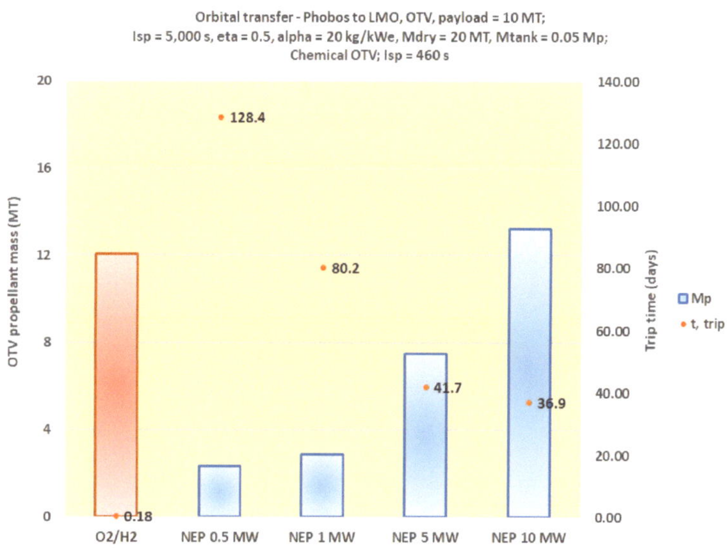

Figure 10.
Resupply propellant mass and round-trip time for O_2/H_2 and Xe Ion NEP OTVs—Phobos to LMO, 10 MT payload mass.

would carry an astronaut crew to the surface of Mars and return to orbit, all with a single stage. The Mars sortie vehicle would be refueled with oxygen and hydrogen propellants created from in situ water resources from the Martian moons. A water electrolysis factory would be delivered to one of the moons and the water would be wrested from the moon's regolith.

In Reference [15], the Mars sortie vehicle was designed to use 80 MT of O_2/H_2 propellant. The initial mass would be approximately 108 MT.

Figure 14 presents the initial mass and propellant mass for a Mars sortie vehicle. The dry mass fraction, B, is varied from 0.1 to 0.25. In the Reference [15] analysis, the 80 MT propellant load would require a somewhat optimistic B fraction of less than 0.10. Using a B fraction of 0.2, the required propellant mass

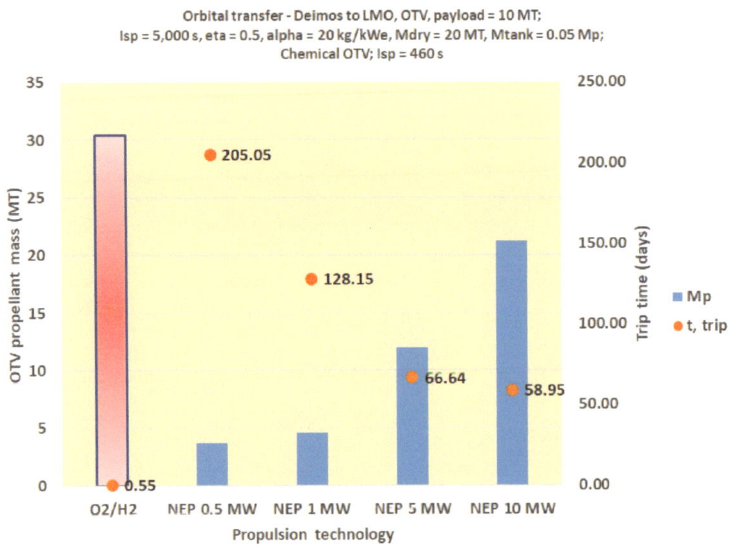

Figure 11.
Resupply propellant mass and round-trip time for O_2/H_2 and Xe Ion NEP OTVs—Deimos to LMO, 10 MT payload mass.

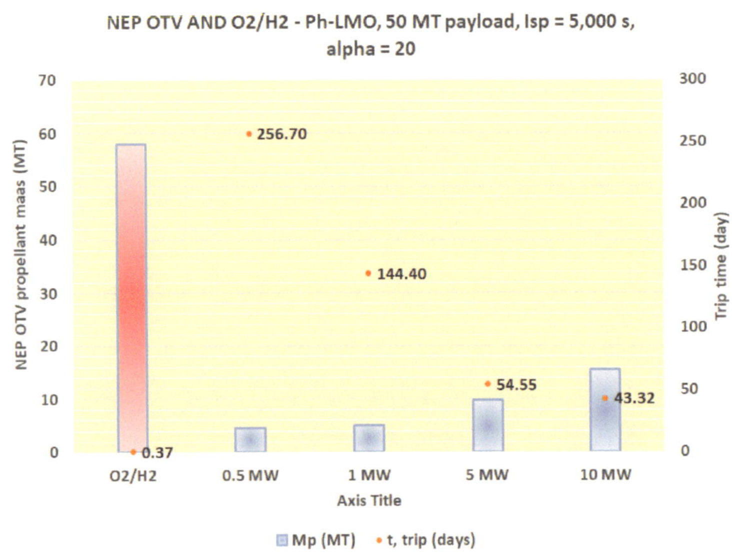

Figure 12.
Resupply propellant mass and round-trip time for O_2/H_2 and Xe Ion NEP OTVs—Phobos to LMO, 50 MT payload mass.

is nearly 200 MT. Therefore five, 40 MT ISRU water resupply flights would be required to support any Mars sortie vehicles.

Using electric propulsion for the resupply flights would enhance the overall architecture, by significantly reducing the total propellant mass needed for the sortie vehicle refueling. Five NEP resupply flights from Phobos would require 50 MT, whereas nearly 300 MT (approximately 6 times the mass) of O_2/H_2 propellant are needed to transport that propellant to LMO. Many propellant deliver benefits are also gained at lower NEP power levels.

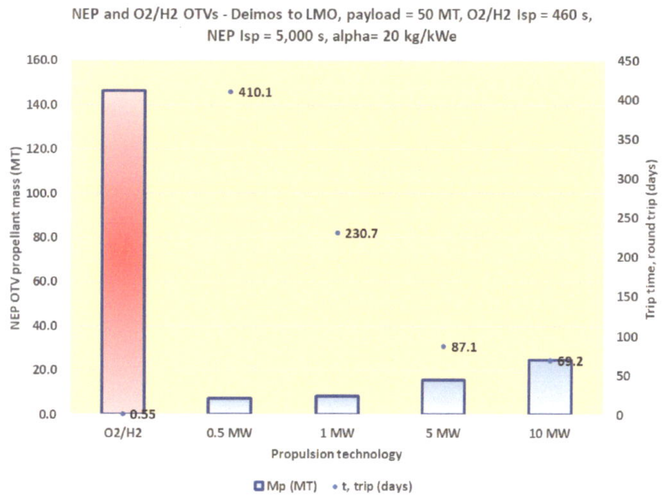

Figure 13.
Resupply propellant mass and round-trip time for O_2/H_2 and Xe Ion NEP OTVs—Deimos to LMO, 50 MT payload mass.

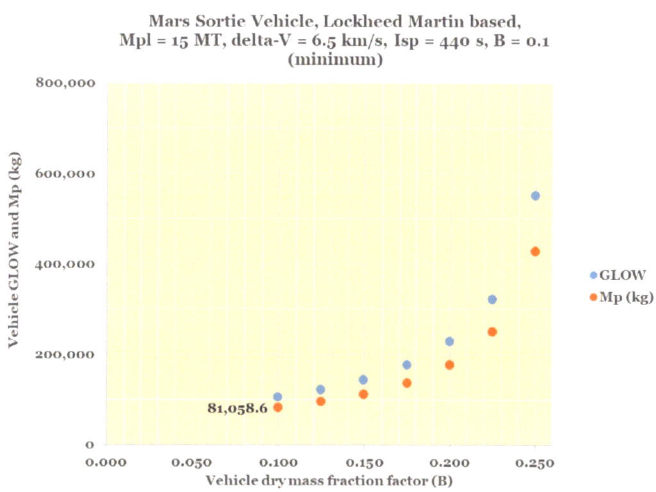

Figure 14.
Mars sortie vehicle initial mass and propellant masses: SSTO capability.

5. Concluding remarks

Electric propulsion offers the ability to transfer large payloads between the Martian moons and in Mars orbit space over O_2/H_2 propulsion. The benefits of electric propulsion are not only in the reduction of propellant masses, but the capability of the high-power reactor system to perform unique science investigations, using radars and other high-energy science instruments.

The NEP systems have a flexible design and can allow many payloads to be manifested together, reducing the overall propulsion architecture. High inclination Mars orbits can be more easily accessed with NEP OTVs with small amount of propellant (compared to the propellant for O_2/H_2 OTVs).

Mining the moons will require specialized factories and processes. The extremely low gravity on the Martian moons will be a challenge in controlling dust, anchoring the spacecraft factory and controlling processes. An artificial gravity factory will likely be needed to maintain the water and propellant processing quality.

Nomenclature

a	acceleration of gravity
delta-V	velocity change
g	gravity level (compared to Earth)
GLOW	gross liftoff weight
H_2	hydrogen
ISRU	in situ resource utilization
LMO	low Mars orbit
m, p	propellant mass
m, pl	payload mass
O_2/H_2	oxygen/hydrogen
SSTO	single stage to orbit

Appendix A: Phobos and Deimos

Phobos
https://nssdc.gsfc.nasa.gov/imgcat/html/object_page/vo1_357a64.html

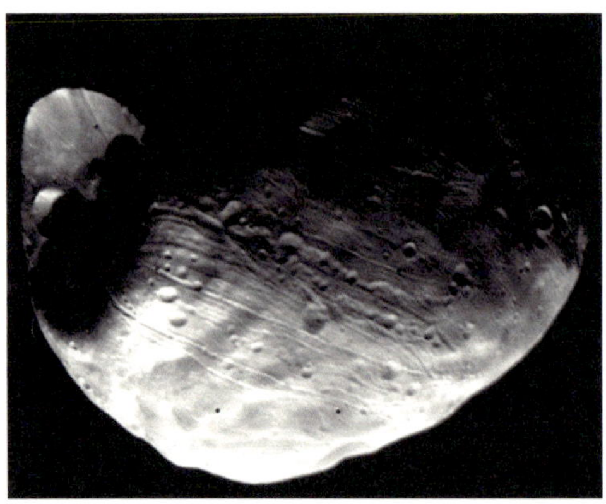

Phobos

Deimos
https://photojournal.jpl.nasa.gov/catalog/PIA11826

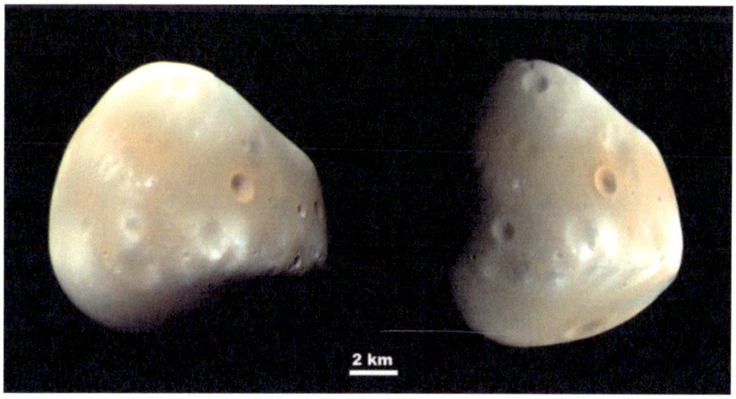

Author details

Bryan Palaszewski
NASA Glenn Research Center, Cleveland, Ohio, USA

*Address all correspondence to: bryan.a.palaszewski@nasa.gov

References

[1] Sheehan W. The Planet Mars: A History of Observation & Discovery. Tucson, Arizona: University of Arizona Press; 1996

[2] National Space Science Data Center. Mars Fact Sheet; 15 June 2020. Available from: nssdc.gsfc.nasa.gov/planetary/factsheet/marsfact.html

[3] Veverka J, Noland M, Sagan C, Pollack J, Quam L, Tucker R, et al. A Mariner 9 Atlas of the Moons of Mars. Icarus. 1974;**23**(2):206-289

[4] Godard B, Budnik F, Muñoz P, Morley T, Janarthanan V. Orbit Determination of Rosetta around Comet 67P/Churyumov-Gerasimenko. 2015. Available from: https://issfd.org/2015/files/downloads/papers/124_Godard.pdf

[5] Godard B, Budnik F, Muñoz P, et al. Orbit determination of Rosetta around comet 67P/Churyumov-Gerasimenko. In: Proceedings of 25th International Symposium on Space Flight Dynamics, Munich. 2015

[6] Pascal L et al. First International Conference on the Exploration of Phobos and Deimos, LPI Contribution Number 1377, November 5-7, 2007. NASA Ames Research Center. Lunar and Planetary Institute; 2007

[7] Bosanac N et al. Manned sample return mission to Phobos: A technology demonstration for human exploration of Mars. In: 2014 IEEE Aerospace Conference. Institute of Electrical and Electronics Engineers (IEEE); June 2014

[8] William S. A Study of System Requirements for Phobos/Deimos Missions. Volume II, Phase I Results—Satellite Rendezvous and Landing Missions. NASA Contract NAS1-10873, Martin Marietta Corp.; 1972

[9] Abercromby A et al. Human exploration of Phobos. In: 2015 IEEE Aerospace Conference. Institute of Electrical and Electronics Engineers (IEEE); 7-14 March 2015

[10] Lee P et al. First International Conference on the Exploration of Phobos and Deimos: The Science, Robotic Reconnaissance, nd Human Exploration of the Two Moons of Mars, Moffett Field, California, November 5-7, 2007. USA: Lunar and Planetary Institute, LPI Contribution No. 1377; 2007

[11] Utashima M. Design and analysis of Phobos-rendezvous orbit. Journal of the Japan Society for Aeronautical and Space Sciences. 1996;**44**(506). Systems Engineering Department, National Space Development Agency of Japan. Published by The Japan Society for Aeronautical and Space Sciences, National Space Development Agency of Japan, NASDA English Translation, December 1997

[12] Palaszewski B. Atmospheric mining of the outer solar system: Resource processing Moon base propulsion, and vehicle design issues. AIAA. 2019. Available from: https://doi.org/10.2514/6.2019-4031

[13] Bate RR, Mueller DD, White JE. Fundamentals of astrodynamics. In: Aeronautical Engineering. Mineola, New York: Dover Publications; 1971

[14] Edelbaum TN. Propulsion requirements for controllable satellites. ARS Journal. 1961;**31**:1079-1089

[15] Cichan T, O'Dell S, Richey D, Bailey SA, Burch A. Mars base camp updates and new concepts. In: IAC-17,A5,2,7,x40817, 68th International Astronautical Congress (IAC), Adelaide, Australia, 25-29 September 2017. Paris, France: Lockheed Martin Corporation; 2017

Section 2

In Situ Operations

Chapter 5

Aerodynamics of Mars 2020 Rover Wind Sensors

Rafael Bardera, Suthyvann Sor and Adelaida García-Magariño

Abstract

Environmental factors in Mars atmosphere are a part of the research issues of the future Mars 2020 mission. The new rover surface vehicle will transport different instruments to investigate the geology, biology, and meteorology of Mars. Amongst these instruments, the Mars Environmental Dynamics Analyzer (MEDA) will be dedicated to the measurement of environment parameters. Two wind sensors will be included in the meteorological station MEDA because wind plays a very important role in Martian climate. High-quality wind data are required to build mathematical models of the Mars climate; therefore, powerful techniques are necessary to eliminate aerodynamic perturbations produced by the rover presence over wind measurements. This chapter is dedicated to the characterization of the aerodynamics around the Mars 2020 rover and its interaction with the rover Mars surface vehicle in order to get information to correct wind data coming from Mars.

Keywords: aerodynamics, rover, Mars, wind, sensors

1. Introduction

The Mars 2020 rover mission is a part of NASA's Mars Exploration Program. This mission is conceived for the exploration of Mars, and additionally, it provides a way to demonstrate novel technologies addressed to the future Martian human expeditions [1]. The investigation of the environmental factors is an overriding aspect to get insight and a better understanding of the meteorological processes in Mars atmosphere. Satellites orbiting Mars provide remote sensing data for the study of Martian atmosphere, but a higher resolution data must be obtained by means of investigation surface vehicles.

The Mars Environmental Dynamics Analyzer (MEDA) is the contribution of Spain to the Mars Exploration Program, and it was designed as a mobile environmental station to be transported by Mars 2020 rover. MEDA sensors will provide information about both ambient and ground, such as wind speed and its direction, temperature, pressure, relative humidity, ultraviolet radiation, and size and shape of dust. MEDA wind sensors are inspired in the design of the wind sensor of Rover Environmental Monitoring Station (REMS) [2, 3] that was embarked on the Mars Science Laboratory (MSL) Curiosity rover [4].

The importance of studying the wind in Mars is manifold. On one hand, in situ propulsion will be necessary for long-term planetary surface missions, and the wind energy could be the power to push wind-driven craft for exploration. On the other hand, quality measurements of the wind in Mars could provide a better understanding of the geophysical phenomena occurring in Mars such as dust devils, carving intracrater layered deposits, and changes on dunes or any other eolian processes due to the wind-driven particle mobility.

2. The Mars atmosphere

The Martian atmosphere is mainly composed of CO2 (95.3%), with small amounts of Nitrogen and other gases. Typical surface atmosphere conditions are a pressure of 700 Pa [5] and a surface temperature variable in the range from 145 to 245 K [6]. The atmosphere density ρ can be calculated considering the perfect gases law as follows:

$$\rho_{Mars} = \frac{P}{R_g T} = \frac{700}{188.918 \times 195} = 0.019\ kg/m^3 \quad (1)$$

where R_g is the constant of CO2 gas taken as $RCO2$ = 188.918 J/kg·K, and T is the mean temperature in Kelvin degrees. The result indicates that the Mars atmosphere density is very low compared with the Earth atmosphere ($\rho_{Earth} = 65\rho_{Mars}$). When the density of a flow is very low, as is this case, usually, the hypothesis of continuum media is investigated by means of the nondimensional Knudsen number (Kn), given by,

$$Kn = \frac{\lambda}{L} \quad (2)$$

where λ is the mean free path, related with the averaged distance between nearest molecules of the gas, and L is a characteristic length scale of the fluidic system or process under study. Although the mean free path depends on the temperature and pressure conditions, a typical value for Mars atmosphere is of order of 10 μm. On the other hand, the typical length of the process can be taken as the characteristic length of rover vehicle ($L \sim 1.7$ m). When analyzing the local flow over rover wind sensors, the boom diameter ($L \sim 0.05$ m) must be taken, resulting in last case, a Knudsen number of order of $2 \times 10^{-4} < < 0.1$, [7] and consequently, the gas of Mars atmosphere can be considered as a continuous media.

On the other hand, the dynamic viscosity (μ) in standard Mars conditions can be computed following the Sutherland law [8]:

$$\mu(T) = \mu_0 \left(\frac{T}{T_0}\right)^{3/2} \left(\frac{T_0 + S}{T + S}\right) \quad (3)$$

where T_0 = 273 K, S = 222 K, and μ_0 is the dynamic viscosity at T_0, taken as $\mu_0 = 1.37 \times 10^{-5}$ N·s/m^2. Computation of Eq. (3) gives a dynamic viscosity coefficient value $\mu = 9.817 \times 10^{-6}\text{N} \cdot s/m^2$ when temperature is 195 K.

Finally, kinematic viscosity coefficient (ν) is given by the ratio of the dynamic viscosity to gas density ($\nu = \mu/\rho$) resulting a value of 5.167×10^{-4} m^2/s. **Table 1** shows the main Mars atmosphere parameters compared with Earth values, including the Mars gravity being 3.7 m/s^2.

From the aerodynamics point of view, the most important parameter to evaluate incompressible flows is the Reynolds number (Re), which is a dimensionless parameter defined as the ratio of inertial forces ($\sim \rho V^2 L$) to viscous forces ($\sim \mu V/L^2$) and also indicates how turbulent the flow is. The expression of Reynolds number is as follows:

parameter	symbol	units	Mars	Earth
Pressure	P	Pa	700	101300
Temperature	T	K	195	288
Density	ρ	kg/m³	0,019	1,225
Dynamic viscosity	μ	N·s/m²	9,82×10⁻⁶	1,79 ×10⁻⁵
Kinematic viscosity	ν	m²/s	5,17×10⁻⁴	1,46 ×10⁻⁵
Gravity	g	m/s²	3,7	9,8

Table 1.
Typical atmosphere flow conditions in Mars compared with Earth.

$$Re = \frac{VL}{\nu} \tag{4}$$

where V is the flow velocity, L a typical length, and ν the kinematic viscosity coefficient of the fluid.

Additionally, Reynolds number is a fundamental parameter that satisfies the dynamic similarity, which is necessary to maintain the similarity between both aerodynamic flows, in Mars and in Earth. The Reynolds ratio is given by

$$\frac{Re_{Earth}}{Re_{Mars}} = \frac{V_{Earth}L_{Earth}}{V_{Mars}L_{Mars}}\frac{\nu_{Mars}}{\nu_{Earth}}. \tag{5}$$

Thereby, when a wind is blowing, in both planets, with the same velocity V over the same object with a characteristic length L, different Reynolds numbers are obtained, and the Reynolds ratio only depends on the physical atmosphere properties, such as kinematic viscosity, which is related to thermodynamics conditions (P and T) so that, in this case, the Reynolds similarity law is conversely proportional to the corresponding kinematic viscosity coefficient

$$\frac{Re_{Earth}}{Re_{Mars}} = \frac{\nu_{Mars}}{\nu_{Earth}} = 35.4. \tag{6}$$

Consequently, the Reynolds number in Earth is approximately 35 times higher than these to be expected in Mars surface, when Earth conditions corresponds to standard atmospheric conditions; therefore, Reynolds in Earth only can be equal when using same gas and same thermodynamics conditions as those present in Mars. But, if we are trying to simulate the problem by wind tunnel testing in Earth atmospheric conditions, flow velocity and length in Earth will be reduced in the same quantity of kinematic viscosity ratio, in order to verify the same Reynolds number in both planets. Because velocity can be slightly reduced, but length can be modified in a large range, this way will be followed by wind tunnel testing. For example, an experiment of the same velocity requires a subscaled model of 1/35th scale, while a reduction to 1/2 of flow velocity requires a reduction of approximately 1/18th in all dimensions of the rover model to be tested in Earth.

3. The Mars 2020 rover

Mars 2020 rover is a nuclear-powered ground vehicle that transports a set of scientific instruments dedicated to investigate the Martian surface. The rover is

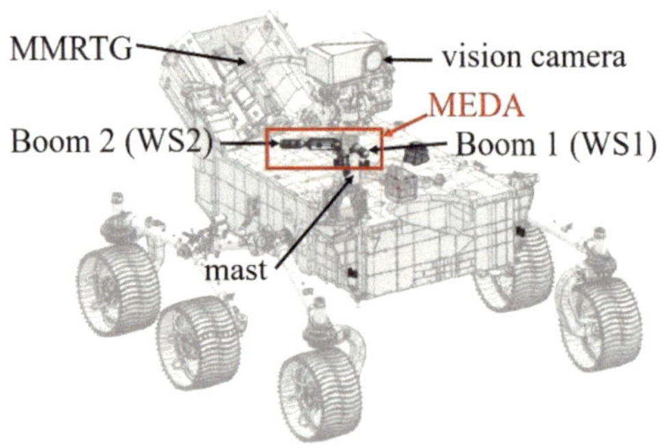

Figure 1.
Scheme of the Mars rover vehicle.

basically composed of a central box body with rectangular section supported by six wheels. A vertical mast supports the remote vision camera and MEDA environmental instruments and an articulated robotic arm is dedicated to get biologic and geologic samples from soil.

Electrical power for engineering systems and science payload of the rover is provided by the Multi-Mission Radioisotope Thermoelectric Generator (MMRGT). The MMRGT converts heat from the natural radioactive plutonium into electrical power [1]. MMRGT is a cylindrical box connected to heat dissipation fins and two heat exchanger plates [9]. **Figure 1** shows a general view of the Mars 2020 rover vehicle (robotic arm was not represented for a better vision of the rover devices).

MEDA wind sensors are located in both booms, around the Remote Sensing Mast (RSM) and oriented 120° from each other. Sensors are capable to measure wind speed up to 70 m/s, and they are located on the rover mast and affected by the rover presence. A correction of the wind data is necessary in order to obtain correct measurements by avoiding the flow perturbation produced by the rover.

4. Thermal anemometry

Thermal anemometry, usually named as hot wire (HW) technique, is a well-established measurement technique introduced in the first half of twentieth century from the study carried out by King after investigating the thin cylinders heat transfer. This technique has got a very high state of development, offering high quality in measuring flow velocities and especially for study of turbulent flows. Air flow velocity and fluctuations can be measured by HW based on the detection of rapid changes in the transferred heat from a tiny sensor (wire typically 5 μm in diameter) to the flow. This sensor is electrically heated so that the flow velocity is determined from the electric power necessary to maintain the temperature of wire at a constant value, by means of an electronic circuit that provides an automatic control loop of temperature. This type of HW is named as Constant Temperature Anemometer (CTA). When electric current crosses through the wire, its temperature is rising due to the heat produced by Joule effect. In a stationary situation, the heat produced is equilibrated by the refrigeration effect produced by the gas flow motion. The energy variation of the wire is given by the following expression:

$$\frac{dE}{dt} = \frac{dW}{dt} + \frac{dQ}{dt} \tag{7}$$

where E is the energy of the wire, Q is the transferred heat to the flow, and W is the heating power, related with the electric intensity I and the resistance of the wire R, given by

$$\frac{dW}{dt} = I^2 R \tag{8}$$

On the other hand, the convicted heat is given by the Newton's cooling law as follows:

$$\frac{dQ}{dt} = h(T_w - T_\infty)A_w \tag{9}$$

where h is the thermal convection coefficient, T_w is the wire temperature, T_∞ is the flow temperature, and A_w is the area of heat transfer for the wire.

Dimensional analysis indicates that thermal convection coefficient depends on the conductivity of fluid k, wire diameter d, and the Nusselt number so that

$$h = \frac{k}{d} Nu \tag{10}$$

where Nu is the nondimensional Nusselt number that represents the convection heat transfer to the conduction heat transfer ratio.

Assuming the orientation of the wire as constant, low airspeed, natural convection is neglectable, and Prandtl number as constant as usual in gases flows, the Nusselt number is only a function of the Reynolds number Re

$$Nu = Nu(Re). \tag{11}$$

This function was empirically determined by King, as the following potential correlation:

$$Nu = C + D\,Re^n. \tag{12}$$

So that, the output signal measured between the ends of the wire is the square of voltage E_W^2 related with wind velocity as follows:

$$E_W^2 = I^2 R^2 = (C_1 + D_1 U^n)(T_w - T_\infty) \tag{13}$$

where coefficients C_1, D_1, and n must be experimentally determined by calibration.

Concluding, the voltage E_W is a measurement directly related with the flow velocity.

5. Rover wind sensors

Several wind sensors are described in the literature [8]. Some sensors based on flow dynamic pressure (Pitot-static probe, rotating sensors as cup and vane anemometers, etc.) were discarded to be used in Mars, mainly due to the very low density of Martian atmosphere, and others as Laser Doppler Anemometry (LDA) because they require an appropriate particle seeding of the flow. Finally, thermal anemometry was selected because it offers reliability in a wide range of temperature, simplicity and robustness, low power consumption, high resolution, and short time of response [4]. Moreover, this technology was used during the Viking mission, demonstrating their measurement capabilities in the Martian atmosphere [10, 11].

MEDA wind sensors are a new design based on the wind REMS sensor [2, 3] that was embarked on the Mars Science Laboratory (MSL) Curiosity rover [4] currently sending data from Mars atmosphere. The behavior of thermal wind sensors installed over the boom of the Mars rover is similar to hot wire, described in earlier paragraph. They are based on a group of four sensitive surface hot film sensors (named dices A, B, C, and D) installed in the same plane (see **Figure 2**). Sensors are refrigerated when the wind is blowing over each dice, and consequently, an electric power must be apply to keep temperature constant on the dice (constant operation reference temperature CTA). The value of electric power gives a measurement of the airspeed on each dice and the direction from blowing. Dices situated in front the flow are cooled more than those located in back positions, since the wind heats up as the first ones are cooled and heat is transferred to the airstream. A comparative analysis of the electric power required for each of the four dices will give us an indication of the angle of incidence of the air flow.

Finally, a joint calibration of the four dices allows us to know the wind speed and the incidence angle over the sensor. A more detailed description is in Ref. [4].

Figure 3 shows the set-up for functional tests of rover wind sensors. A simplified version of hot film sensor, similar to the real sensor was implemented over a flat plate. A specially designed bed fabrication in extruded polystyrene was used to support the plate with hot film sensors in order to perform the functional tests by wind tunnel testing to verify the wind endurance. During this test campaign, several values of airspeed were blown by wind tunnel and the air was flowing on the sensors refrigerating them. Readings of electric signals from dices electronic chip were acquired to verify the correct behavior of the wind sensors, and its integrity was verified when the flow was blowing over them. A laser Doppler anemometer was used as an airspeed standard.

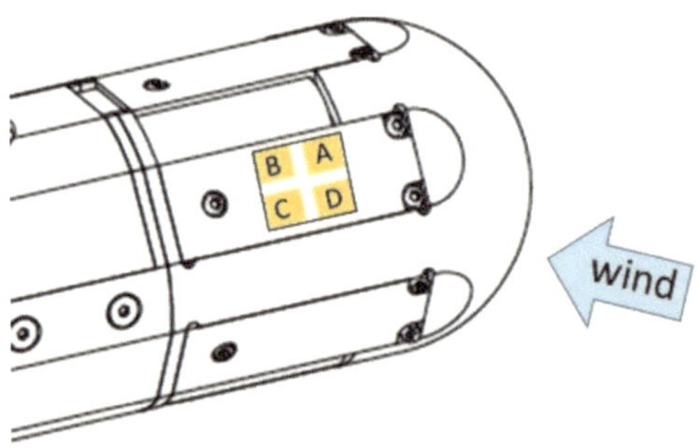

Figure 2.
Rover wind sensor boom.

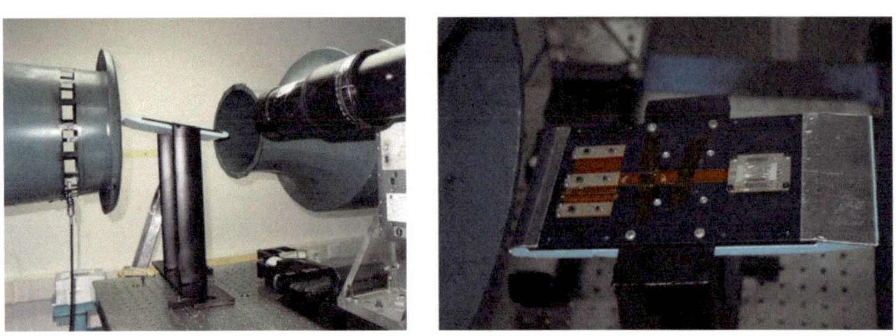

Figure 3.
Mars rover wind sensors during functional tests.

6. Boom aerodynamics

The rover wind sensors are installed on the boom surface. Booms are located in the vertical mast over the upper rover central box surface (see **Figures 1** and **2**). Two booms are at the same height and contained in the same horizontal plane, with an angular offset of 120°. The boom external geometry is similar to these of the Pitot-static tube. This is a classical instrument for measuring airspeed in aerodynamics and fluid mechanics that has proven its efficiency for more than a century. This tube has a hemispherical head followed by a cylindrical body with axisymmetric geometry so that the longitudinal flow is coming to the wind sensor without detachment. On the other hand, rover booms have a hemispherical head followed by a cylindrical body with polygonal faces where thermal sensors are installed (see **Figure 2**). In this section, a theoretical analysis of two simplified cases is presented: longitudinal axisymmetric flow and cross flow over the boom.

6.1 Longitudinal axisymmetric flow

When the wind is blowing in longitudinal axisymmetric direction, the head of the boom receives the flow without detachment. A first approximation to this kind of flow can be studied by considering the incompressible potential flow on the axisymmetric Rankine half-body of revolution. This flow is produced when a uniform stream of velocity U_∞ is flowing over a three-dimensional point source of strength M located at the origin. This situation is depicted in **Figure 4**, for a half-body of revolution with radius a.

A stagnation point denoted by P is produced by the upstream source point when both singularities are equal in velocity. The location of this point is calculated as follows:

$$U_\infty = \frac{M}{4\pi d^2}, \tag{14}$$

and finally, the distance to point P is

$$d = \frac{1}{2}\sqrt{\frac{M}{\pi U_\infty}}. \tag{15}$$

The velocity potential Φ corresponding to the superposition of a uniform stream and a source located in the origin is given by the following expression:

$$\Phi(\mathrm{x}, \mathrm{r}) = U_\infty x - \frac{M}{4\pi R} \tag{16}$$

where $R = \sqrt{x^2 + r^2}$.

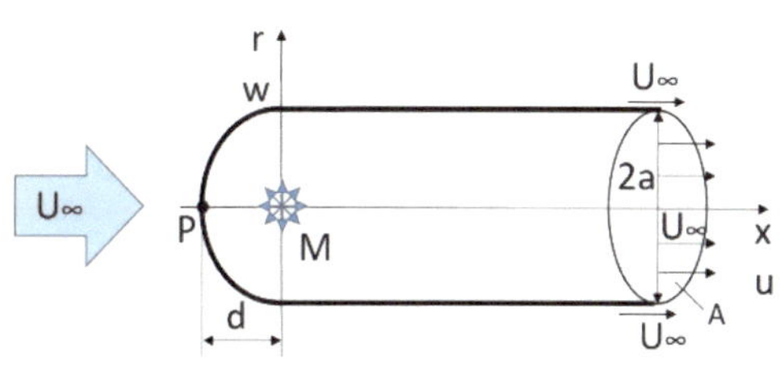

Figure 4.
Axial flow over wind sensor boom.

On the other hand, the Stokes stream function Ψ for this flow is given by,

$$\Psi(r,z) = \frac{M}{2}\left(1 - \frac{x}{R}\right) + \pi r^2 U_\infty. \tag{17}$$

The half-body surface stream function is equal to the flow rate coming from the inner of the half-body. Taking into account that it is a body of revolution, transversal area is $A = \pi a^2$ and the flow coming from the inner body, crossing the circular area denoted as A (see **Figure 4**), is AU_∞, exactly the strength of the point source ($M = \pi a^2 U_\infty$), and the distance d is computed being $d = a/2$.

The dividing streamline represents the surface body, and it is obtained when the stream function is equal to the source strength, M,

$$M = \frac{M}{2}\left(1 - \frac{x}{R}\right) + \pi r^2 U_\infty. \tag{18}$$

And the corresponding velocity is

$$u(x,r) = \frac{\partial \Phi(x,r)}{\partial x} = U_\infty + \frac{M}{4\pi} \frac{x}{(x^2 + r^2)^{3/2}} \tag{19}$$

$$w(x,r) = \frac{\partial \Phi(x,r)}{\partial r} = \frac{M}{4\pi} \frac{r}{(x^2 + r^2)^{3/2}}. \tag{20}$$

Velocity over body surface is computed taking into account that both variables x and r are related by the body stream function as follows:

$$\frac{M}{2}\left(1 + \frac{x}{R}\right) = \pi r^2 U_\infty \tag{21}$$

Resulting in

$$r^2 = \frac{a^2}{2}\left(1 + \frac{x}{\sqrt{x^2 + r^2}}\right) \tag{22}$$

Taking nondimensional variables defined as

$$\chi = \frac{x}{2a} \tag{23}$$

$$\zeta = \frac{r}{2a}. \tag{24}$$

Both variables are related by the stream function as follows:

$$8\zeta^2 = 1 + \frac{\chi}{\sqrt{\chi^2 + \zeta^2}}. \tag{25}$$

The variable χ results in a function of ζ, given by,

$$\chi^2 = \frac{\left(8\zeta^2 - 1\right)^2}{16\left(1 - 4\zeta^2\right)}. \tag{26}$$

Finally, the longitudinal velocity component is given by the following expression:

$$\frac{u(x,r)}{U_\infty} = 1 + \frac{1}{16}\frac{\chi}{\sqrt{(\chi^2+\zeta^2)^3}}. \tag{27}$$

And simplifying, we can find an expression that only depends on the unique variable, ζ, as follows:

$$\frac{u(x,r)}{U_\infty} = 1 + \frac{(8\zeta^2-1)\sqrt{(1-4\zeta^2)}}{64}. \tag{28}$$

Figure 5 shows the graph of nondimensional velocity u/U_∞ versus the nondimensional distance $x/2a$. The velocity is growing from the stagnation point located at $\chi_P = -1/4$ until unity nondimensional velocity that is reached when curve cross the origin ($\chi = 0$). After that, a peak is produced, and the freestream velocity value is got for a longitudinal distance $x/2a = 1.5$. So that, when the flow is coming to a boom section located at a distance of approximately equal to three times the radius ($x = 3a$), the wind flows with the freestream velocity, and the hot film sensor will be refrigerated by the same manner as in nonperturbed conditions.

6.2 Transversal flow

Wind flowing transversal to the boom (boom in cross-flow) can be studied as a first approximation by the two-dimensional complex potential flow over a circular cylinder, as represented in **Figure 6**.

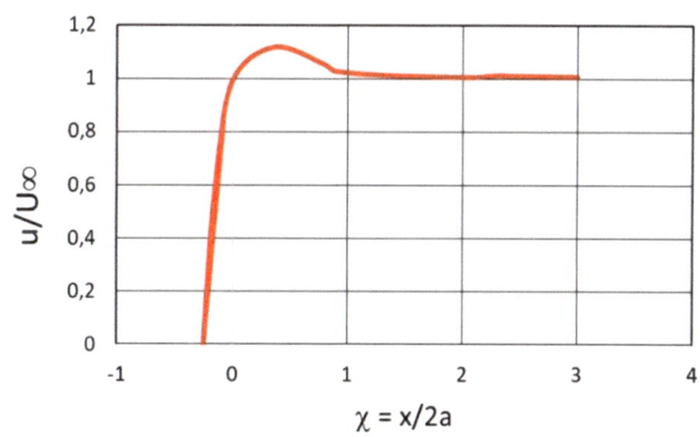

Figure 5.
Nondimensional wind velocity on rover boom in axial flow.

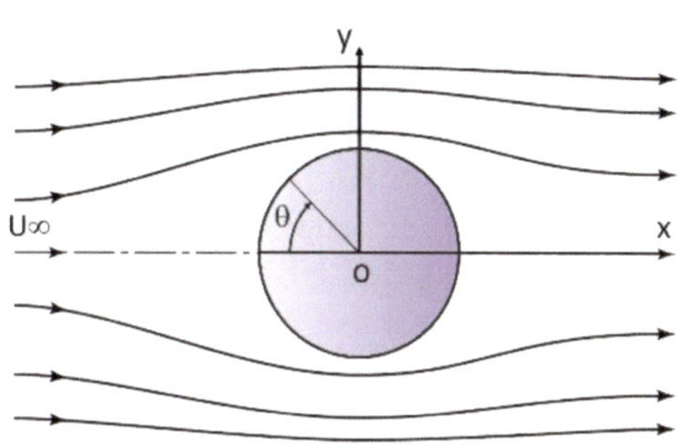

Figure 6.
Potential flow of a cylinder.

The complex potential $w(z)$ corresponding to a cylinder with radius a is given by the superposition of a uniform stream and a doublet [12]

$$w(z) = U_\infty \left(z + \frac{a^2}{z} \right) \tag{29}$$

where z is the complex variable defines as follows:

$$z = x + iy. \tag{30}$$

Real and imaginary parts of w define the potential of velocity ϕ and the stream function ψ, respectively,

$$\phi = U_\infty \left(r + \frac{a^2}{r} \right) cos\theta \tag{31}$$

$$\psi = U_\infty \left(r - \frac{a^2}{r} \right) sin\theta. \tag{32}$$

The conjugate velocity is obtained directly by derivation as follows:

$$\frac{dw}{dz} = U_\infty \left(1 - \frac{a^2}{z^2} \right) = u - iv. \tag{33}$$

This expression can be simplified by making use of the Euler formula $z^{-2} = a^{-2}e^{-i2\theta} = a^{-2}(\cos 2\theta - i\ \sin 2\theta)$, and the velocity components over the cylinder surface are the followings:

$$u = 2U_\infty \sin^2\theta \tag{34}$$

$$v = -2U_\infty sin\theta cos\theta. \tag{35}$$

The square of velocity magnitude is given by

$$|V|^2 = u^2 + v^2 \tag{36}$$

$$|V|^2 = U_\infty^2 \left[(1 - \cos 2\theta)^2 + (\sin 2\theta)^2 \right] \tag{37}$$

$$|V|^2 = 2U_\infty^2 (1 - \cos 2\theta). \tag{38}$$

And finally, the modulus of velocity over the cylinder surface is

$$|V| = 2\, U_\infty sin\theta. \tag{39}$$

When the angle is zero, the velocity is zero, corresponding to a stagnation point, but when angle is 90°, the velocity is $2\, U_\infty$, rising to the maximum value of velocity.

The potential flow approximation is valid only in face front the flow, but it is not valid in the wake because the flow is detached. **Figure** 7 shows a comparison view between attached flow upstream cylinder (left) and detached flow behind the cylinder (right).

6.3 Boundary layer

Potential flow is used as a first estimation of flow parameters, but it is not a real flow because viscosity is not present in these flow models. The no-slip condition

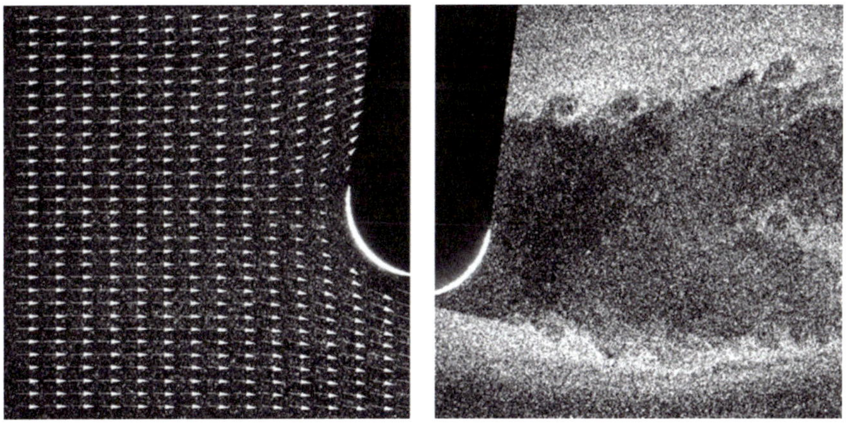

Figure 7.
Real flow around a circular cylinder: Upstream cylinder (left) and behind the cylinder (right).

over the body wall needs zero velocity in the wall and the outer velocity (U_∞) in the boundary layer limit. The boundary layer is a thin layer where the effects of viscous forces are of the same order of the inertial forces. The Reynolds number of flow outer boundary layer is larger than unity, and consequently, the inertial forces are larger than the viscous. When the wind is flowing on the boom surface, the boundary layer is growing downstream and the turbulence and adverse pressure gradients produce the separation of the boundary layer and the detachment of flow, as it can be observed in **Figure 8**. In this situation, wind is flowing over booms in very complex conditions to be simulated analytically and the flow coming to each side of boom depends on the incidence angle of wind, so that thermal sensors are operating in a no ideal condition and heat transfer mechanisms are far from a theoretical environment. It is because the booms are calibrated isolated in order to characterize the output electrical signals from sensor when they are subjected to different wind conditions.

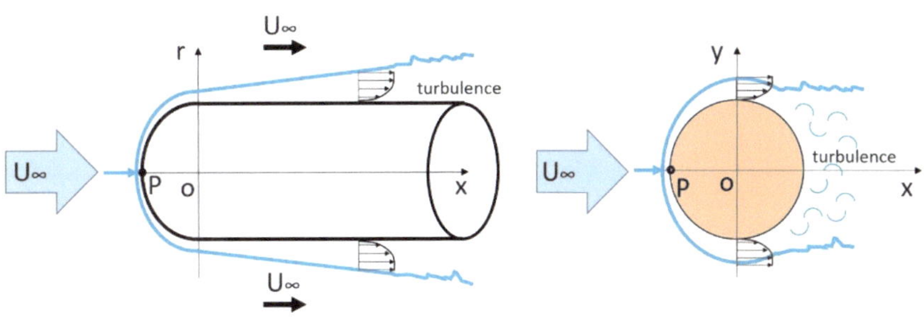

Figure 8.
Boundary layer over the boom.

7. Rover vehicle aerodynamics

The external surface of rover is not designed following aerodynamics criteria. The main body of this vehicle is a box with rectangular section, supported by six wheels. A vertical mast is erected on the upper surface of the main box body. The instruments of MEDA are installed over this mast. Two booms are located perpendicular to the axis mast. The rover vehicle is larger than other elements as mast and booms, and the flow viewed by small devices is mainly affected by the presence of the rover.

The flow visualization is a qualitative technique that offers a global view of the flow so that it provides a first approximation to the study of the flow. **Figure 9** shows the smoke visualization at low Reynolds number when the flow was illuminated by a laser light sheet. This experiment was performed in a small commercial wind tunnel from TSI (model 8390) installed in our laboratory. It was running at an airspeed of 2.6 m/s. This tunnel has a square cross test section of 100 × 100 mm^2 with transparent walls, in order to provide optical access for the recording camera. The smoke was produced by burning incense bars.

The flow is coming from right to left. Photo in left-hand side shows the streamlines of flow when curved by the presence of the mast and the camera, and a detached flow is produced downstream. On the other hand, photo in right-hand side shows the flow when coming from rear of the rover. The flow is deviated by effects of a large box that corresponds to the module of MMRGT and a detached stream is produced behind this module. Only some streamlines are coming to the camera height but curved.

Figure 10 shows the smoke visualization carried out in the INTA-1 wind tunnel. This facility has maximum airspeed of 60 m/s and closed circuit with open test section of 2 × 3 m^2. The flow around the rover was running at high Reynolds number and it was illuminated by conventional white light. The flow is steered to the rover presence forming a fine smoke tube, but the direction of the smoke is slightly deflected as it is coming to the rover vehicle rising the boom 1 with an angle different to zero as corresponds to no disturbed horizontal flow, which is the effect of the presence of the rover. The direction of the wind received by sensors is changed by the rover presence. Flow is coming from right to left.

The quantitative analysis of the flow was performed by means of Particle Image Velocimetry (PIV). This is a nonintrusive technique that illuminates the flow seeded with tracer particles and provides simultaneously, the flow field velocity in a plane of the flow [13].

Figure 9.
Smoke visualization of flow around rover (Re = 6770).

Figure 10.
Smoke visualization of flow around rover (Re = 2.2 × 10^5).

Figure 11 shows the averaged velocity map in a vertical plane of the flow containing the boom 1 of the rover, when the flow was blowing with zero angle of incidence and airspeed of 10 m/s coming from left to right. The color scale represents the velocity magnitude in meter per second. The deflection of streamlines can be observed near the boom 1, although the flow is attached. On the other hand, the flow in the wake of the mast is detached and enclosed in a low-velocity recirculation bubble.

Figure 12 shows the averaged velocity map in a horizontal plane of the flow containing both booms 1 and 2, because they are installed at the same height over the mast. No symmetric deflection of streamlines is observed. This effect is due to the presence of the rover central box and the vertical mast, which is located in the right side of the rover. Flow is coming from left to right and the mast wake is clearly visible, with low velocities (blue color). The wind sensor 2 (WS2) is located as an appendix forming 120° with camera axis.

The flow over the rover was investigated at low Reynolds number (6770) by wind tunnel testing a small rover model with a characteristic length of 38 mm (scale 1: 45th). A TSI model 8390 wind tunnel specially adapted to very low flow velocities was used (mean airspeed of 2.6 m/s). This tunnel has a square cross test section of

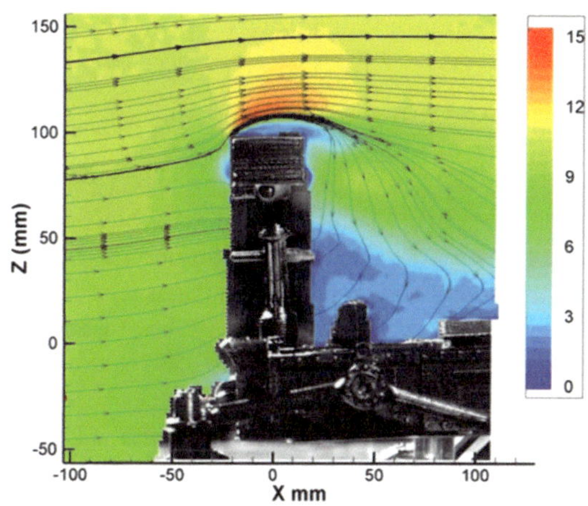

Figure 11.
Velocity map in a vertical plane (Re = 2.2 × 10^5).

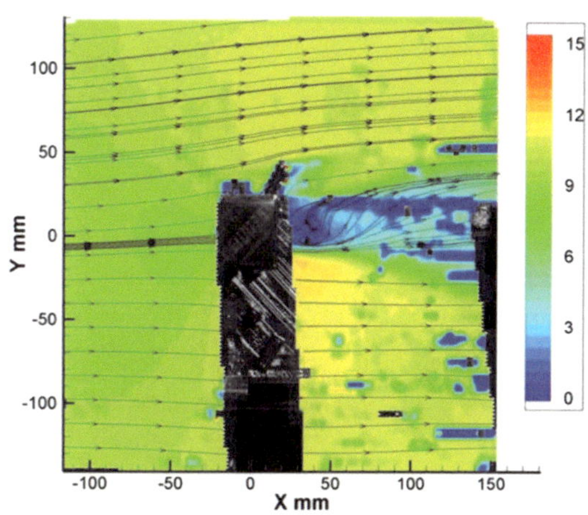

Figure 12.
Top view: Velocity map of a horizontal plane (Re = 2.2 × 10^5).

100 × 100 mm^2 with transparent walls of methyl methacrylate. It provides optical access for experimental techniques and visual access to the experimentalist.

Laser Doppler Anemometry (LDA) is a punctual nonintrusive measurement technique that is usually used to investigate the flow field velocity with very high accuracy and resolution [14].

Figure 13 shows the experimental set-up, with a body axes system fixed to the model. The angle of wind incidence is defined as β, and the location of both booms is indicated, and the orientation of the LDA probe remains fixed while the model is turning on to simulate different incident wind angles.

LDA measurements were carried out by using a commercial system from Dantec. The illumination source consists of a 10 mW He-Ne laser beam emitting in red color (wavelength of 632.8 nm). A BSA-F60 Flow Processor and BSA Flow software provide data of flow velocity. Seeding particles were produced by means of a water ultrasonic atomizer. **Figure 14** shows the rover model inside the wind tunnel test section during the test experiments and measuring velocity by LDA. Rover vehicle was supported by a circular plate that facilitates the rover turn on during test campaign. Reflexions of red LDA laser beam can be observed in the walls of wind tunnel test chamber.

Modulus of wind velocity in sensor 1 location is determined from three orthogonal velocity components measured by LDA after the following expression:

$$V_1 = \sqrt{U_{LDA}^2 + V_{LDA}^2 + W_{LDA}^2}. \tag{40}$$

An analogous expression was used to compute the velocity at the location of sensor 2.

Figure 15 shows the modulus of velocity as seen by both wind sensors WS1 and WS2. Differences with freestream velocity are observed in wind velocity detected by laser Doppler in respective locations of rover wind sensors. Velocities were measured for different incidence wind angles in range 0–360°. Two forbidden bands that correspond to not valid data are marked in **Figure 15**. These areas correspond to detached flow when the sensors are located mainly in the wake of camera mast and MMRGT of rover. In practice, the solution is to use active operating redundancy, since both sensors are operating simultaneously in order to complement the forbidden regions, so that whole angular sector can be measured by rover sensors.

The sensor 1 (WS1) is the main sensor because this sensor will be used for wind measurements except in the not valid region. The sensor 2, named as WS2, must be used in the forbidden band of sensor 1.

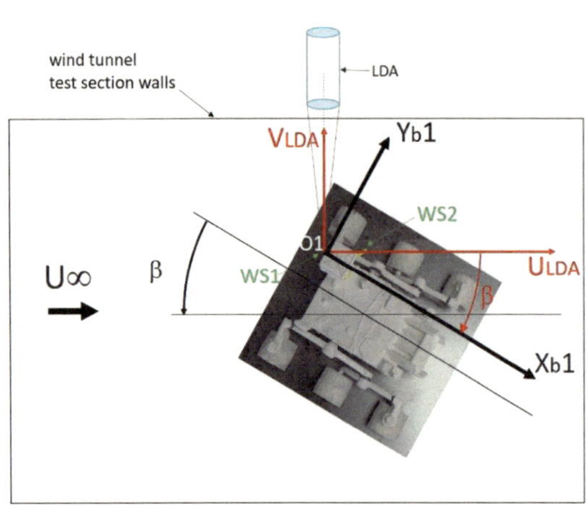

Figure 13.
Top view: Velocity map of a horizontal.

Figure 14.
Rover model inside the test section during LDA measurements.

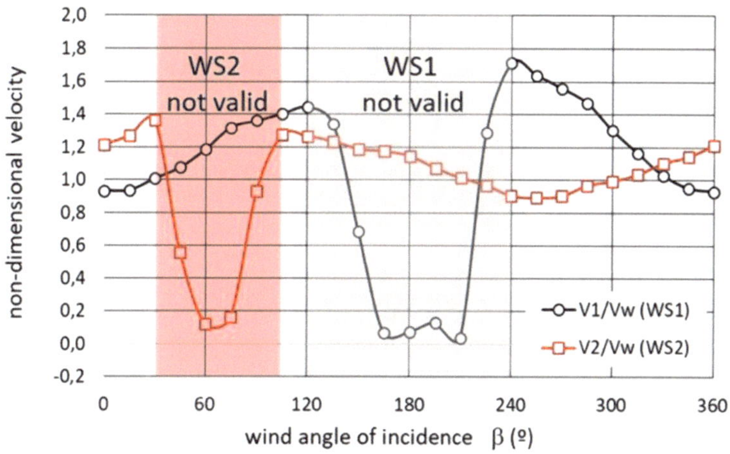

Figure 15.
Wind velocity measured by rover booms.

On the other hand, measurements provided by both sensors in the cut-off angles (30, 105, 120, and 240°) are valid, and the wind velocity in these points would be determined by an averaged velocity value.

Figure 16 shows forbidden circular sectors for both wind sensors measurements. When wind is blowing from forbidden angles of incidence, the flow is detached by a combined effect of the camera mast and MMRGT of rover that produce a turbulence flow in the mast wake that is coming to the booms where the rover wind sensors are installed. Graphs in **Figure 16** demonstrate that complete circular angles range can be measured by the combination of both wind sensors.

Finally, the velocity measured by each wind sensor must be corrected from the effects of the external geometry of the rover vehicle. The true wind velocity V_W is given by

$$V_W = K_i V_i; i = 1, 2 \tag{41}$$

where K_i is the calibration factor that must be determined for each sensor, from data contained in **Figure 15**, by computing calibration factors by the following expression:

$$K_i = \frac{V_W}{V_i}. \tag{42}$$

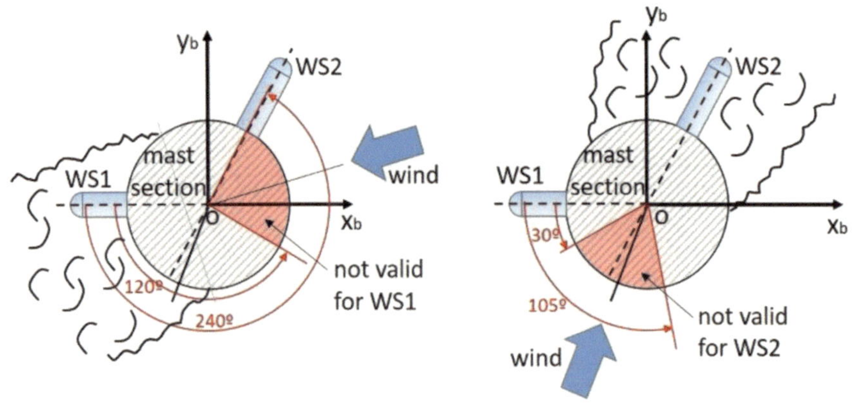

Figure 16.
Forbidden circular sectors for wind sensors measurements.

This ratio is precisely the inverse of this presented in **Figure 15** and it represents the effect of rover vehicle over the booms.

8. Conclusions

The Mars 2020 rover aerodynamics plays a fundamental role in the operation of rover wind sensors. A set of hot film sensors have been specially designed to be installed over two booms located perpendicular to camera mast of rover. This kind of sensors are based on the heat transfer from sensors to environment around the sensors, but physical properties of Earth atmosphere are very different to these present in Mars atmosphere where sensors will operate during Mars 2020 mission.

Reynolds number is the fundamental parameter to establish the physical similarity between real flow in Mars and this investigated in Earth by wind tunnel testing.

Potential flow provides a first approximation to study the longitudinal and transversal flows over booms where wind sensors will be installed. Limitations of the potential model are evidenced by detachment flow in the wake of booms with transversal flow and the presence of the boundary layer where viscous effects are of the same order of inertial effects. Due to these limitations, a more intensive experimental investigation is necessary. Global information about the flow pattern around the rover vehicle was obtained by smoke visualization. Qualitative techniques as Particle Image Velocimetry (PIV) and Laser Doppler Anemometry (LDA) were used to get insight about the velocity field. The effects of rover vehicle over the flow coming to the sensors booms were studied, and not valid regions of each sensor were determined.

Finally, the calibration factor equation was indicated as the main way to correct flow velocity from the adverse effects produced by the rover vehicle over wind sensors measurements.

Acknowledgements

This investigation was funded by Spanish Ministry of Defense under the program “IGB 99001 Termofluidodinámica” of INTA (National Institute for Aerospace Technology of Spain).

Nomenclature

A	transversal area of body boom
A_w	area of heat transfer for the wire
a	radius of the boom
C	empirical coefficient
C_1	empirical coefficient
D	empirical coefficient
D_1	empirical coefficient
d	wire diameter
E	energy of wire
E_W^2	voltage of the wire
g	gravity
h	thermal convection coefficient
I	electric intensity crossing the wire
k	conductivity of fluid
K_i	calibration factor of velocity
Kn	Knudsen number
L	characteristic length scale
M	strength of a three-dimensional point source
n	empirical coefficient
Nu	nondimensional Nusselt number
Q	transferred heat to the flow
R	electric resistance of the wire
Re	Reynolds number
$RCO2$	constant of CO2 gas
R_g	constant of gas fluid
S	reference temperature
T_w	wire temperature
T_∞	flow temperature
u	flow velocity component
U_∞	freestream velocity
U_{LDA}	velocity component measured by LDA after x axis of rover
v	flow velocity component
V	flow velocity
$\lvert V \rvert$	modulus of velocity
V_{LDA}	velocity component measured by LDA after y axis of rover
V_W	true wind velocity
V_1	wind velocity measured in location of boom 1.
W	heating power of wire
W_{LDA}	velocity component measured by LDA after z axis of rover

Greek symbols

λ	mean free path
ρ	density
Φ	velocity potential
μ	dynamic viscosity coefficient of the fluid
ν	kinematic viscosity coefficient of the fluid
Ψ	stream function

Author details

Rafael Bardera*, Suthyvann Sor and Adelaida García-Magariño
National Institute for Aerospace Technology, Madrid, Spain

*Address all correspondence to: barderar@inta.es

References

[1] Tahu G. Environmental Impact Statement for the Mars 2020 Mission. NASA Reptort 20546; Washington, D.C.; 2014

[2] Gómez-Elvira J, REMS Team. Environmental monitoring instrument for Mars exploration. Lunar and Planetary Science. March 10-14, 2008. pp. 1647-1648

[3] Gómez-Elvira J et al. REMS: The environmental sensor suite for the Mars science laboratory rover. Space Science Reviews. 2012;**170**:583-640. Springer

[4] Domínguez M, Jiménez V, Ricart J, Kowalski L, Torres J, Navarro S, et al. A Hot Film Anemometer for the Martian Atmosphere. Planetary and Space Science. USA: Elsevier; 2008. pp. 1169-1179

[5] Petrosyan A, et al. The Martian Atmospheric Boundary Layer, Reviews of Geophysics, 49, RG-3005/2011. American Geophysics Union

[6] Leovy CB. Martian meteorology. Annual Review of Astronomy and Astrophysics. 1979;**17**:387-413. Annual reviews Inc

[7] Schetz JA, Fuhs AE. Chapter 11: Rarified gas dynamics. In: Handbook of Fluid Dynamics and Fluid Machinery. USA: John Wiley & Sons, Inc; 1996. p. 672

[8] White F. Fluid Mechanics. Fourth ed. Boston, Massachusetts, USA: McGraw-Hill Companies, Inc.; 2002

[9] Bhandari P, Anderson K. CFD analysis for assessing the effect of wind on the thermal control of the Mars Science Laboratory Curiosity Rover. In: International Conference on Environmental Systems (ICES). AIAA, Vail, CO; 2013

[10] Chamberlain TE. Atmospheric measurements on Mars: The Viking meteorology experiment. Bulletin American Meteorological Society. 1976; **57**(9):1094-1104

[11] Hess SL, Henry RM, Kuettner J, Leovy CB, Ryan JA. Meteorology Experiments: The Viking Mars Lander, Icarus. Vol. 16. USA: Elsevier, Inc.; 1972. pp. 196-204

[12] Kundu PK, Cohen IM. Fluid Mechanics. Second ed. USA: Elsevier Science; 2002. pp. 160-163

[13] Raffel M, Willert CE, Wereley ST, Kompenhans J. Particle Image Velocimetry: A Practical Guide. Berlin, Heidelberg, Germany: Springer-Verlag; 2007

[14] Albrech HE, Borys M, Damaschke N, Tropea C. Laser Doppler and Phase Doppler Measurement Techniques. New York, NY, USA: Springer; 2003

Chapter 6

Evolution of the Scientific Instrumentation for *In Situ* Mars Exploration

Andoni G. Moral Inza and Guillermo Lopez-Reyes

Abstract

Mars has always been a magnet for the human curiosity. The more we know about the red planet and its past, the more complex are the unanswered questions. In order to answer them, an ambitious long-term plan for the robotic and manned exploration of Mars has been established by the scientific community worldwide. To ensure success in answering the issues to be investigated on each step of the plan, the selection of "on board" payloads at mission level is specifically designed for achieving the best possible results. This selection also has modified the mission operation modes from a set of individual experiments to a cooperative science paradigm where all the instruments in the mission payload contribute jointly to achieve unprecedented scientific results. Collaboration not only between experiments but also between agencies for achieving major goals has been demonstrated as the optimum way forward for Mars exploration. This chapter presents a historical review, with a look into the future, of the human efforts aimed at understanding the red planet, focusing on the technological advances and scientific discoveries achieved that help answer some of the most thrilling and transcendental questions ever raised by humanity: Are we alone in the Universe?

Keywords: Mars, rover, lander, *in situ* instrumentation, collaborative science

1. Introduction

The study of ancient Mars, as well as its evolution to the planet we can observe today, has become one of the major scientific and technical challenges in the field of planetary exploration. Knowing that Mars was once covered with liquid water, that it had quite an intense geological and volcanic activity, together with the presence of a much denser and rich atmosphere suggest the existence of conditions compatible with the appearance of life, also quite similar to primitive Earth. Considering the vast knowledge we already have about Mars (though probably still a tiny fraction of what is there to learn!), and its accessibility and proximity as our neighbor in the solar system, the study of the red planet is still today the best candidate to look for existing or extant life, or at least to find traces that life might have emerged or existed in those long gone ancient favorable conditions.

It is true that recent studies and discoveries related to the icy moons of the gas giants of the Solar System might present most favorable conditions for past, but also present, existence of life. The evidence of the presence of water liquid oceans below the icy crust of Europa, or the powerful magnetic field that protects the surface of Ganymede, or Io's intense volcanic activity encourages the idea of considering these Jovian moons as potential life reservoirs. This is also true for the moons of Saturn Titan, with a dense methane atmosphere and a great deal of complex organic molecules; and Enceladus, whose surface is also covered by a thick ice layer. Nevertheless, the technical complexities to reach these planetary bodies, together with very long mission duration make these objectives unreachable today. We will have to wait for the technology and political will to evolve during the following years to ensure the viability of *in situ* missions to these moons in the decades to come.

With these considerations, it is quite clear how the exploration of the Martian surface and shallow subsurface has emerged as the best chance at answering some of the most fundamental questions of humankind: Are we an exceptional fortuity in the Universe, in the galaxy, in the Solar System? Does the appearance of life in our planet answers to a natural process in the evolution of the Universe? In this case, if life or traces of life were to be found in Mars, where there were no plate tectonics and which we know had similar characteristics to primitive Earth, we would be able to study the primal forms of life, forever lost in the ever-recycling Earth mantle.

In this context, this chapter will guide the reader through the evolution of the different payloads used for the exploration of Mars since the kickoff of the space career in the 1960s, providing insights to the technological advances and scientific discoveries achieved, bringing the reader into a time travel into the future of the *in situ* exploration of Mars.

2. The pioneering missions

The timeline of the missions to Mars reflects the evolution of the technical capabilities throughout the history of planetary exploration. This technical evolution, together with the understanding of the planet by previous missions, has helped shaping the increasing complexity of the scientific questions investigated: observation from Earth, fly-bys, orbiters, landers, rovers, etc., which will be followed by helicopters (Mars 2020) and subsurface exploration (ExoMars 2022), missions aimed for the return of Martian samples and finally human exploration and, who knows? maybe the establishment of permanent bases on Mars. Every reached milestone, as will happen with those still to come, allowed to contrast and validate scientific hypothesis formulated about Mars formation and evolution; and also open the door to those questions to be answered by missions to come, especially concerning current habitability hazards.

The pioneering missions acquired great momentum with the Moon space race of the 1960s between the US and the USSR, becoming a renewed Mars competition. As a result, every launch window to reach the red planet became exploited as an attempt to flyby first and then to reach its orbit. Every piece of information or picture about its surface or atmosphere and ambient conditions was considered fundamental for planting the flag first. The race between the Mariner and Марс projects had started.

2.1 First contact with Mars: Mariner 4 to 7

Though the USSR launched Марс 1 in 1963, the first probe ever sent to flyby another planet, it lost the communication when it was at more than 100 million kilometers away from the Earth. This failure cleared the path for the first ever mission ever to fly above Mars to photograph it, the Mariner 4 (1965). This probe, equipped with a TV camera threw down all the theories regarding the presence of liquid water on the Martian surface (the astronomers' observations of canals on the late nineteenth and early twentieth centuries were still quite present in the collective memory), confirming the terrestrial observations: a desertic and rocky surface with craters, no oceans, and a very thin atmosphere [1, 2]. However, the images also showed more geologically interesting features than expected. Also surprisingly, the magnetometer onboard the ship did not detect any magnetic field [3]. This, together with the lack of a radiation belt around Mars as confirmed by the trapped radiation detector (TRP) instrument [4], showed that the Marian surface was exposed to the solar radiation without any protection.

These results were surprising at the time, and were of course relevant from an astrobiological perspective, as the preservation of organics and/or life tracers on the Martian surface is surely impaired by the intense and unfiltered radiation. These data were complemented by Mariner 6 and 7, identical probes launched during the 1969 launch window. The aim of these ships was the study of the Martian surface and atmosphere (without any measurements during the cruise phase). To fulfill these objectives, a new instrumentation package was developed, also paving the way for future missions. The Mariner 6 and 7 payloads were included in addition to the TV camera, spectrometers in the IR and UV ranges, and an IR radiometer.

The results of these missions showed a topographically complex surface, where not only the craters reported by Mariner 4 were present but also distinctive topographic forms were observed: chaotic and featureless terrains [5]. These kinds of structures need active modification processes to occur, contrary to a Moon-like crater-only landscape which would imply inactivity since the very old ages. In addition, it was observed how in extreme latitudes ice layers formed in the rim of craters, evidencing water activity on the planet.

With Mariner, infrared spectroscopy showed powerful capabilities as it was used to analyze a wide range of parameters on Mars: polar caps, surface and atmospheric composition, or surface temperature and topography. The results showed that the Martian atmosphere was composed by CO_2 with traces of water vapor [6]. On the surface, IR spectrometry detected goethite, an oxidized chemical compound associated with weathering processes in the presence of water. This confirmed for the first time the possibility of a wet ancient Mars. Also, the surface temperature was measured by the IR radiometer, showing values around 140 K [7].

2.2 Visiting the planet: first orbits and soft landing

Traveling a stable orbit around Mars was a milestone reached virtually at the same time by Mapc 2 and 3, and Mariner 9, all of them launched in 1971. Even if Mapc 2 and 3 departed some days before Mariner 9, a faster cruise phase of the latter allowed it to be the first space probe to orbit another planet by a margin of two weeks. However, the main scientific objectives of Mariner 9 were to continue with the studies of the Martian atmosphere started by Mariner 6 and 7, while mapping the Martian surface. Profiting from the relatively low distance orbit (1600 km, the closest at the time), together with a Visual Imaging System consisting of up to nine cameras with notably better resolution than previous missions (98 vs. 790 m per pixel), Mariner 9 was intended to map 70% of the Martian surface during its mission.

At their arrival to Mars, the probes were greeted by a great sandstorm which lasted for several months. This was of course unforeseeable and had a severe impact in the missions. The soviet orbiters were the most affected, as they were mostly able to photograph the sand clouds above the surface, but the worse was still to come since these missions also included two landers that were liberated for landing as programmed, suffering the consequences of the storm: Mapc 2 crushed against the Martian surface, while Mapc 3 could certify the first soft landing on the surface of Mars. However, this was a bitter success, as it could only operate during 20 s before (probably) the storm made it lose communications. Mapc 3 was an extremely ambitious mission (probably too much at the time), as it included a small rover, Prop-M, connected with an umbilical cord to the lander platform. The early failure of the mission made impossible to know if the Passability Estimating Vehicle for Mars was successfully deployed on Mars. It took 25 years for a rover to be successfully deployed on the Martian surface.

Instead of that, the US Mariner 9 mission was just an orbiter, but had however one critical advantage compared to their competitors: an onboard patchable software during the mission. This became a mission-saver for Mariner 9, and a space-race win for the US, as the mission ground control modified the plan to save resources during the storm duration and observe the Martian moons in the meanwhile. Once it settled down, Mariner 9 started mapping the Martian surface, sending back to Earth more than 7000 images covering practically 100% of the planet surface. These images showed river basins, huge ancient volcanos, very long canyons, etc., together with evidences of erosion phenomena caused by water and wind [8]. This mapping, together with the confirmation and more precise study of the Martian atmosphere density and pressure, or the surface temperatures with the infrared radiometer (IRR) instrument, allowed the compilation of all the necessary information to prepare, with the maximum possible confidence, the future landing of the Viking missions.

3. Start of the surface missions: Viking program

As the United States started showing a position of dominance in the space race, the urge for committing to launches at every opportunity relaxed (also due to a reduced economic impulse). This way, the 1973 launch window was not used, and the first Viking launch occurred during the summer of 1975. This program was born with three objectives: acquire high-resolution images from the Martian surfaces; continue with the surface and atmosphere chemical analysis; and to look for evidences of life on the Martian surface. Viking was also conceived as a twin mission (similar to Mapc 2 and 3), each of them with an orbiter plus a lander.

3.1 The Viking orbiters

The main objective of the Viking orbiters was to help on the selection of the landing sites for the landers, as well as serve as communication relays. This was central to the mission, considering the lessons learned from the Soviet Mapc 2 and 3 failures, which could not select the landing site, and needed direct contact with Earth for operation. During more than 1 month, the information gathered by the landers was used to localize and certify the best possible locations to perform the soft landing.

However, the orbiters were also equipped with their own payload, which was reduced compared to their Mariner predecessors, as the lander was onboard. This payload included the IR Mars atmospheric water detector (MAWD), to study the presence of atmospheric water vapor and its latitude and seasonal potential variations [9, 10]. The infrared thermal mapper (IRTM) instrument measured the surface temperature, confirming the night/day cycle temperature variations, as well as characterizing its variability associated with latitude, seasons and atmosphere [11]. Finally, the visual imaging subsystem (VIS) included two high-resolution television cameras with a 17 m/pixel resolution. The geologic analysis of these samples supported the theory that vast ancient liquid water surfaces were present on the surface of Mars. Not only hydro fluvial systems but also geological features compatible with lakes or other water reservoirs were related to weathering process throughout the planet [12].

3.2 The Viking landers

The landers started their mission the moment they were released from the orbiters. During their descent, information regarding the composition, structure, and temperature of the planet ionosphere was obtained. Furthermore, the UAMS mass spectrometer analyzed the higher layers of the atmosphere, while the lander monitored the atmosphere pressure and temperature along the descent. But of course, the leap forward by the Viking landers was the success of sending the first-ever images of the Martian surface after a soft landing, setting a new milestone in the technological development for planetary exploration. This way, on July 20, 1976, Viking 1 landed on the western area of Chryse Planitia (22.27 deg N latitude and 312.05 deg E longitude); and her twin landed on September 3, 1975, 200 km west from the Mie crater in Utopia Planitia (47.6673 deg N latitude, 134.2809 deg E longitude).

The 600 kg Viking landers were equipped with a very complete suite of experiments to try to reach the ambitious objectives planned for the mission: the analysis of the surface and ambient properties derived from the erosion and eolian sedimentations; morphology, organic, and inorganic chemical composition and magnetic properties, based on the mineralogical analysis of the landing site; seismology; meteorology; and to look for potential Martian organisms with a biological experiment attached to a gas chromatographer/mass spectrometer (GC/MS) instrument.

The panoramic cameras on the landers covered a region of 360° of the Martian horizon, but also allowed photographing the lander and its sample-extraction arm, as well as the sun or the Martian moons, Phobos and Deimos. These cameras were the first to be operated, starting to transmit the first data to ground only after 25 s. The 3000 images of Viking 1 plus the 6500 of Viking 2 showed a desertic, powdery, and inhospitable landscape. Also, the images greatly helped in the interpretation of the instrumental data.

The temperature and magnetism of the rock samples in the reach of the landers were analyzed by sensors placed in the arm tip, showing a great abundance of magnetic minerals in the Martian surface [13]. The X-ray fluorescence spectroscopy (XRFS) were used to obtain the chemical nature of the surface regolith, showing a great abundance of Si and Fe, with significative concentrations of Mg, Al, S, Ca, and Ti. When compared with the abundances of these elements on Earth, it was observed that the presence of S was up to two orders of magnitude higher in Mars, while K abundance was 5 times lower than the average found in the Earth crust [14].

The Viking meteorological station was deployed in a mast after landing and included temperature and wind sensors. It also included a pressure sensor under the belly of the lander. All these instruments gathered data as configured from ground, varying the data logging along the mission as required. The results showed and allowed the characterization of the day/night cycles, as well as throughout the seasons during the years the missions lasted [15].

One of the Martian unknowns was related to the seismic activity of the red planet. The Viking landers seismometers were included for this reason. Even the Viking 2 instrument failed (probably due to the landing impact); the seismometer on Viking 1 worked for 2000 h without registering any important events, with the exception of one, which could have been caused by a micro-meteorite impact or even a wind gust [16].

Finally, the key to the mission was the Viking experiment aimed at answering the question of whether Mars might have harbored life in the past or even if there was life present in the soil. In order to do so, Viking included the "biological experiment" consisting of three different instruments: a pyrolitic release (PR), labeled release (LR), and gas exchange (GEX). All these instruments incubated samples extracted from the Martian surface, applying different ambient conditions during several days. In general, the search for organic traces were negative in five out of the six experiments. The remaining one, however, performed by the LR instrument on Viking 2, obtained a positive result [17]. Revolutionary at the beginning, this result was always surrounded by controversy regarding the goodness of the experiment, resulting in serious doubts on the existence of this positive biological response. This controversy is still in force, though the general scientific position nowadays lean the scales toward a false positive result. Even if with controversial or unsatisfactory results, the Viking missions were a great success for the study of Mars, also leaving very important lessons learnt for the generations to come, especially when coming down to looking for alien life or life-tracers. On the one hand, habitable is not the same as inhabited; on the other hand, positively certifying the existence of life requires either repetitive results, and/or a very good assurance that the organic/biological detection is coming from the extraterrestrial source, in order to avoid potential controversies on the results. This might be one reason why, since 1976, no Mars mission has been intended to look for life on Mars, but to look for conditions of habitability.

4. Mobile surface missions: from Pathfinder to Curiosity

Twenty years had to pass before the world looked again toward Mars after the successful Viking missions. The emotional slump caused by the difficulties in finding life on Mars, together with the slowing down of the space race due to the clear leadership of the US after the Viking program, was among the causes of this lack of interest. However, after this time, NASA recovered the impulse considering the new geopolitical scenario in which the study of the Solar System was not anymore a race between countries, but a joint effort among all of them. This way, Europe, Japan, and even Russia were considered as potential allies in this new era of exploration. It is in this framework that the NASA Mars Exploration Program (MEP) [18] kicks-off in 1993. This program laid an ambitious strategy for 1993–2020, with the idea of using all the available launch windows with subsequent missions to study Mars, its climate and geology, available resources for *in situ* exploitation, and the search for life. The original path planned a (now we now) quite optimistically sample-return missions for the decade of 2000, continuing with manned missions to the red planet before 2020.

Understanding the evolution of our neighbor Mars, with which Earth shares a common geological origin, and a similar habitability in their origins 4000 million years ago, is the main objective of the MEP program, the greatest effort for the exploration of the Solar System since the Apollo missions. One of the MEP key characteristics has been the focus on the development of technologies that provide security and reliability on the Martian missions. This has resulted in a high success rate, which has pumped an increasingly great and continuous scientific understanding of Mars. A timeline showing the roadmap for the exploration of Mars can be found in **Figure 1**.

Such an ambitious program requires interdisciplinary collaboration with the Planetary and Martian scientific community, which is done through the Mars Exploration Program Analysis Group (MEPAG)—a scientific group formed by international experts from the main national space agencies around the world; but also MEP interacts closely with the NASA Human Exploration Operation Mission

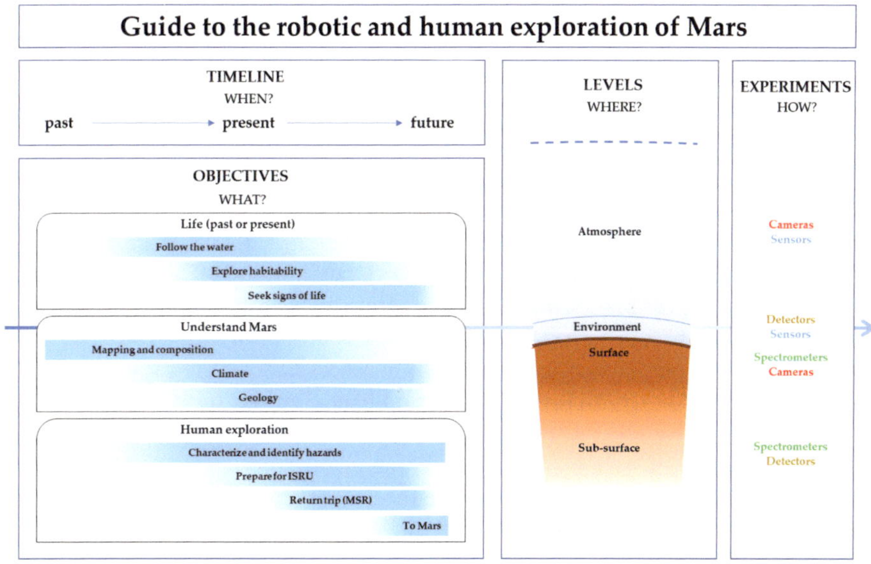

Figure 1.
Key elements and timeline of the robotic and human exploration of Mars.

Directorate (HEMD) and the Space Technology Mission Directorate (STDM), as covered by NASA's 2014 Strategic Plan [19].

As part of this plan, the MEP objectives are updated as necessary with the new scientific discoveries, while considering the following priority aspects: (1) a continued effort for the development of reliable technologies to improve the analytical capabilities with new and more ambitious scientific instruments; (2) a technological evolution to allow for better and safer technologies for entry, descent, and landing (EDL) in order to be able to land more equipment smoother, and with higher confidence and accuracy; and (3) to keep a reliable and continued network of communication relays, by maintaining a sufficient network of orbiters as the best way to facilitate a sustained flow of scientific discoveries around Mars. This approach allows for better instruments, placed with higher accuracy and safety, while ensuring a data flow between Earth and the Martian robots [20].

In order to meet these objectives, MEP has successfully combined orbital and surface missions as needed. On the one hand, orbital missions have provided climate and atmosphere compositional studies, combined with mapping and surface characterization activities to facilitate future landing sites. But they also have served to analyze the planetary characteristics such as magnetic field and solar particle interaction, while fulfilling the communications link with Earth for the surface missions. Successful orbiter missions to date have been the Mars Global Surveyor (MGS) in 1996; Mars Odyssey (ODY) in 2001; Mars Express (MEX), a European mission lead by ESA in 2003; Mars Reconnaissance Orbiter (MRO) in 2005; Maven (MVN) in 2013; and the ExoMars Trace Gas Orbiter (TGO), also from ESA, in 2016.

On the other hand, lander surface missions were proposed to answer maybe more concrete scientific objectives: the Mars Pathfinder (MPF) in 1997, basically a technological demonstrator also equipped with a small rover; Phoenix (PHX) in 2007, which looked for the presence of ice in high latitudes; and Interior Exploration using Seismic Investigations, Geodesy and Heat Transport (InSight) in 2018, mainly dedicated to the study of the planet seismology.

But the true heroes, those missions that have harvested the greatest scientific and public relations success are the exploration rovers. Sojourner is the Mars Pathfinder mission rover in 1997; Spirit and Opportunity are the Mars Exploration

Rovers (MER) from 2003, which surprised the world with amazing science for 15 years (while they were designed to last for a nominal mission of 90 sols); and finally, the Curiosity rover, or Mars Science Laboratory (MSL) in 2012, deploys the most complete analytical laboratory ever on another planetary body beyond Earth.

4.1 First orbiters and Pathfinder; 1996: technological readiness

The first steps of the MEP were not easy. Early years of the program were marked by a number of sounding failures: Mars Observer, in 1992, an orbiter, was NASA's first mission for Martian exploration after the Viking project, and lost communications before entering into orbit; and the Mars Climate Orbiter, in 1998, crashed against the planet surface. But not only NASA had problems, other countries either: The Soviet Union lost Phobos 1 and had a limited success with Phobos 2 in 1988. The Mapc 96 in 1996 was also lost. Japan never arrived to Mars with the Naomi mission in 1999.

Not only the technological issues were a concern during the decade of the 1990s but also the geo-economic situation made it difficult to justify missions with the cost and size of previous decades. This way, the Mars Environmental Survey (MESUR) Pathfinder mission was conceived inside the low-cost planetary Discovery Missions program. This was a new mission concept by NASA grounded in the "faster, better, and cheaper" paradigm. However, an unexpected ally appeared with this mission to gain the popularity and acceptance of the US and world public opinion: Internet. On July 4, 1997, the Pathfinder landing was forecasted through Internet, reaching unprecedented records of visits to the Webpages offering information about the mission. NASA called it "the day Internet stood still." From that experience, NASA understood that in order to win the public opinion favor in order to get acceptance on this new era of planetary exploration, public relations activities advertising all aspects of the missions were key to feed a public avid of this kind of information. And Internet was the perfect means of transmission, allowing almost real-time updates on the missions.

This framework helped restart the exploration of Mars, paving the way for a new generation of planetary exploration missions by the US. The low-cost discovery missions such as Pathfinder had the clear objective of demonstrating the technological readiness to land and explore on Mars, investing only three years and with a low cost: safe landing systems, new communications, the use of modern sensors and image devices, etc., but above all, to demonstrate the capability of maneuvering in the Martian surface with a rover.

The Mars Pathfinder landed softly on Ares Vallis (19.33 N, 33.55 W), on Chryse Planitia (already visited by Viking 1). The lander was named Carl Sagan Memoria Station (after the famous astronomer Carl Sagan); and the rover was called Sojourner (in honor to the American civil rights activist Sojourner Truth). Sojourner was the first rover ever to be successfully deployed outside the Earth-Moon system (after the failure of Mapc 3 in 1971, the Soviet Union sent two rovers to the moon surface later in the decade of the 1970s as part of the Lunokhod program). The surface bi-dimensional planetary exploration era was started.

Even though the mission was basically considered as a technological demonstrator, it included several instruments as part of the platform and rover payloads. The lander's main objective was not only to help on the rover operations but also to work as a meteorological station and to investigate the magnetic properties of the Martian powder. The meteorological observations were done by the Atmospheric Structure Instrument/Meteorology Package (ASI/MET). Deployed on the platform mast, it included several temperature, wind, and pressure sensors. The observations performed during the 80 operative sols of the Pathfinder showed daily pressure variations of 0.2–0.3 mbar, with two complete cycles, correlated with the temperature variations observed during the Martian day. The temperature observed during the equatorial summer ranged from 263 to 197 K in the day/night cycles. The winds also followed daily patterns, being able to detect some dust-devil episodes as well. All the results were in agreement with those measured 21 years before by the Viking 1 [21].

The magnetic analysis of the Martian powder by the Magnetic Properties Investigation (MPI) experiment showed that the detected iron oxides (Fe_2O_3, in its phase gamma-maghemite) could be used to infer their water-related formation. However, a higher-than-expected (compared to Earth environment) abundance of goethite (alpha-FeOOH) and hematite (alpha-Fe_2O_3) was detected by this instrument [22], raising new scientific questions about their formation.

The images acquired by the Pathfinder lander Imager for Mars Pathfinder (IMP) instrument were used to contextualize the images by the Sojourner rover, and these also showed a complex surface with rims and canals clearly weathered by fluvial, eolian, and impact processes. Also, some atmospheric processes such as the formation of clouds were observed [23].

The Sojourner rover was equipped with an Alpha Proton X-Ray Spectrometer (APXS) devoted to analyze the composition and atmospheric and water-related weathering of the surface materials from a geochemical perspective. The results showed rock compositions similar to the ones present in the Earth crust (as well as in the SNC meteorites). Thanks to this, it was inferred that volcanic processes were present in the planet geological history, as mafic components and volcanic gasses are needed for the formation of this kind of rocks [24, 25].

4.2 MER, 2003+2003: follow the water

After the success of the Pathfinder mission in 1996, NASA's Mars Odyssey was successfully set in orbit around Mars in 2001. The next launch window in 2003 placed ESA's Mars Express (MEX) in orbit, but saw the failure of the Beagle 2 lander after being unable to completely deploy its solar panels. This window was also selected by NASA to launch the most ambitious scientific mission to date: the Mars Exploration Rover (MER). This mission was intended to land two twin rovers, Spirit and Opportunity, on opposite points of the planet, in order to answer a set of scientific questions following the motto *follow the water*. The experience and results gained by NASA and the scientific community with the Pathfinder mission marked the path for the exploration of Mars: water was the key.

4.2.1 The mission objectives

The questions that the MER missions were to address could be summarized as follows:

1. **Was life ever present on Mars?** Assuming life as what we know, the presence of stable liquid water is needed for long periods of time on the planet surface. The presence of water, then, increases the probabilities (though it is not necessarily essential) of harboring life. In this sense, and even that the rovers were not equipped to directly detect life even if present, they were capable of performing mineralogical analysis that would help deduce the formation and evolution conditions of the analyzed materials. So, the search for minerals formed or evolved in water-related processes (precipitation, sedimentation, evaporation, hydrothermal processes, etc.) was priority for the MER rovers.

2. **What are the current weather conditions on Mars? What were they like in the past and why did they change?** Again, the mineralogical analysis of the

composition of the Martian rocks and soil can be used to infer the ambient conditions under which they are formed, including potential water-related alteration processes. In addition, the rovers were also designed to study the lower layers of the atmosphere to help understand the current Martian weather.

3. **What are the geological characteristics of Mars? What role the wind, water, plaque tectonics, or volcanism have played in the Martian geological history?** Iron compounds, carbonates, clays, salts, and other minerals in which water plays an important role were candidates to answer this kind of questions. Analysis performed on these samples would also help confirm the measurements performed by orbiter missions such as MGS, Mars Odyssey, Mars Express, or the Mars Reconnaissance Orbiter that improved their detection capabilities with the help of the MER rovers.

4. **Learn how a human exploration mission could be setup.** To study the Martian environment to identify potential risks for humans; and to understand and gather information regarding the chemical nature of the minerals and local resources, to be potentially used for exploitation and use during a manned mission, including course water in any of its forms, as it is the only way to ensure long-term manned missions on the Martian surface.

4.2.2 The robotic geologists

The Spirit and Opportunity rovers were designed with these questions on mind, considering the rovers as geologists *roving* on the Martian surface, and equipped with a portable laboratory to perform *in situ* analysis of the samples. A 360° panoramic camera (Pancam) and an IR thermal emission spectrometer (Mini-TES) were placed on the rover mast at 1.5 m above the ground. These allowed the optical characterization of the terrain (the geologist eyes) plus a first IR analysis on the potentially interesting targets. In addition, the rovers were equipped with an articulated robotic arm with several tools: a Microscopic Imager (MI) to take closeup images of the samples (the geologist magnifier); the Rock Abrasion Tool (RAT) to allow access to the interior of the rocks (the geologist hammer); and a series of instruments constituting the geologist portable laboratory—the Alpha Particle X-Ray Spectrometer (APXS), the Mossbauer Spectrometer (MS), and a Magnet array for direct analysis of the samples (another Magnet array was placed on the rover chassis to passively analyze the powder depositing on them). The rovers were also equipped with navigation (NavCams) and Hazard Avoidance (HazCams) cameras to facilitate a safe roving capability of these robotic geologists.

4.2.3 A journey through the Martian geology

The MER rovers landed on Mars at the beginning of 2004, during the last days of the southern summer. The soft landings were achieved by means of airbags, with which the rovers bounced very long distances along the Martian surface for a period of time much longer than expected. Spirit landed on January 4, 2004 in the Gusev crater (14.572° S, 175.478° E). Three weeks later, on January 25, Opportunity landed inside a 20-m diameter crater in Terra Meridiani (1.946° S and 354.473° E) in the opposite side of the planet.

MER rovers' journey on the Martian surface is one of the greatest successes among all the planetary exploration missions. From an engineering perspective, the results are impressive: rovers were designed for a 90-sol nominal mission and to rove up to 600 m. However, both rovers operated successfully for years (Spirit lasted for 6 years and 2 months, or 1892 sols; and Opportunity worked for 14 years and 42 days, or 5111 sols). During this time, they managed to travel 7.73 km in the case of Spirit, and a stounding distance of 45.16 km in the case of Opportunity. This longevity and traveling capacity allowed the generation of tens of GB of data that were downloaded to Earth. This enormous data volume allowed an unprecedented identification and consolidation of scientific advances to date, contributing greatly to the knowledge of Mars and its geological history.

4.2.3.1 A dense atmosphere and fresh water

First Opportunity success was its initial lucky-shot, landing inside the Eagle crater, where it performed analysis for 2 months. It detected hematite, which was a clear indicator of a past presence of water in the area, as it is formed in water environments. However, it was also deduced that the water presence was salty and with low pH (so not optimal for life thriving due to the water acidity), with this area probably being a coast region with tidal waters [26]. Later in December 2011, Opportunity also detected other evidences of liquid water in the Martian past. In the Endeavor crater, the instruments detected veins of gypsum (calcium sulfate) inside some rocks. This hydrated mineral was probably formed by water flowing through cracks in the rocks, where the calcium was left behind [27]. Before that, in October 2005, while analyzing the Comanche outcrop by MS, APXS and Mini-TES, the Spirit rover identified rocks formed by key chemical elements such as magnesium and iron carbonates, in proportions up to 10 times higher than the previous analysis on Mars [28]. This discovery was evidence that the Martian past had warmer and wetter areas, with a thick atmosphere and with neutral pH values that are required for the formation of these carbonates. These conditions would potentially favor the existence of some kinds of microscopic life compared to any of the previous analysis performed on the Martian surface.

4.2.3.2 Thermal waters

One of the Spirit wheels was damaged long after the nominal mission had passed, which was in itself a stroke of luck. In March 2007, the stuck wheel uncovered a whitish area below the soil regolith, which, after analysis, revealed as an area composed by silica with 90% concentration. This kind of crystals, with this purity, can only be found (on Earth) in areas bathed by thermal waters, or water vapor currents where the water or vapor get in contact with volcanic rocks. These places in Earth are thriving with bacteria and microorganisms due to the optimal conditions set by the hot and wet ambient [29].

4.2.3.3 An active water cycle

The Spirit mission looked terribly bad when it was trapped in sand without any possibility of moving from that position. However, during the maneuvers to try to be released, the wheels uncovered sulfates (among other things) under the regolith in the Troy location. These sulfates seemed to have been in contact with water only 1 million years ago (a very short time by geological standards), suggesting the possibility of the existence of an active water cycle on the planet.

4.2.3.4 Stable water bodies and volcanic activity

In May 2007, Spirit observed the ancient remains of a volcanic eruption in Home Plate. The remains suggested that the explosion might have been caused by a water-lava interaction [30], as suggested by the "bomb sag" structures found in the lower layers of the plateau. These structures are formed on Earth by the rocks falling on soft surfaces, which would confirm the presence of stable water bodies heated up by volcanic activity, which could also be favorable for microscopic life.

4.2.3.5 Ambient and atmospheric conditions

In addition to water-related discoveries, both Spirit and Opportunity helped performing many other discoveries and studies to further the knowledge of the red planet. Opportunity became an expert on the geology of Martian craters [31] after visiting more than 100 impact craters, which allowed understanding their formation and erosion on the Martian atmosphere. The Martian environment was monitored as well, studying the cloud formation and the suspended powder and opacity, and how it affected the solar panels of the rovers (curiously, the next rover sent to Mars would be equipped with a nuclear battery instead of being solar-powered). The dust-devils on the surface (firstly detected by the Pathfinder) were photographed by Spirit, and were key for the mission extension, as they helped recovering power for the rovers by cleaning the powder accumulating on the solar panels [32]. A complete temperature profile of the Martian atmosphere was performed by Opportunity, combining Mini-TES data [33] with the Mars Global Surveyor orbiter TES instrument data. Also, Opportunity found the first extra-Martian meteorite on Mars called Heat Sink Rock [34].

Spirit and Opportunity rovers were decidedly successful missions that provided prolifically scientific evidences of the past presence of water on the Martian surface among other things. Whichever of those discoveries would have absolutely justified the missions by themselves; but considering MER missions' success all together is simply overwhelming.

In order to prepare for the missions to come, NASA launched the Mars Reconnaissance Orbiter (MRO) in 2005, with the objective of mapping and facilitating communications. And the *follow the water* motto would be closed after the limited results obtained by the Phoenix lander in 2008, which landed in the polar regions of Mars (68° N latitude), but did not keep up with the very high expectations of this unexplored area. So, once the existence of an ancient wet Mars had been proved, the next milestone in the Martian exploration was to understand if the water presence could facilitate conditions favorable for the appearance of life: *explore habitability.*

4.3 Curiosity (MSL) 2011: explore habitability—seek for signs of life

The foundations laid by the Mars Exploration Program, with communications guaranteed by several orbiters on Mars (ODY, MEX, and MRO were operative in 2012; with two other on their way: MVN and TGO), and the experience gathered during the MER missions in many aspects, including the power source for the rovers, facilitated a science-centered design for the next rover to be operated on Mars. The Mars Science Laboratory (MSL), or Curiosity rover, landed in 2012 with only a small mass dedicated for communications (an UHF antenna) with orbiter relays. It also incorporated a light radioisotope thermoelectric generator (RTG) as a power source, which would also guarantee a stable power source for the rover (contrary to the solar-based power system on the MER). Furthermore, MSL introduced a new landing method based on the famous *sky crane*, which greatly reduced the mass of the entry, descent, and landing (EDL) system, being this mass allocated for the rover itself. This way, on August 5, 2012, the heaviest scientific mobile platform was deployed on Mars, with almost 900 kg of mass devoted to the exploration of the Martian surface.

4.3.1 Mission objectives

The Curiosity rover, equipped with the most powerful set of scientific instruments ever on another planetary body, was bound to determine if Mars could have harbored life at any time in its past, as well as continue understanding the role played by the water to this end; of course, it was also prepared to study the Martian climate and geology. So, the new exploration paradigm migrated from *follow the water* to *explore habitability*, by studying the chemical and structural properties of the soil and rocks, especially those presenting water-related formation scenarios.

Contrary to MER, Curiosity included instrumentation capable of performing analysis which can be related to biological studies or processes in order to address the following objectives: the search of organic carbon compounds or biosignatures; geological and geochemical analysis based on the analyzed rocks; analysis of the Martian climate and its evolution; and also to prepare potential future manned missions by being able to characterize the planet radiation on the surface.

4.3.2 The rover instrument suite

Lacking from a landing platform, the Curiosity rover incorporated some instruments to help during the EDL stage, such as the Mars Descent Imager (MARDI), or the atmospheric sensor MEDLI (MSL EDL Instrument). Also, the NavCams and HazCams on the rover ensure a safe navigation system for the rover. In addition to these, the rover contains a very complete suite of scientific instruments.

4.3.2.1 Cameras

MASTCAM is a color panoramic camera that is used for macroscopic analysis. It is used to establish the geological context of the analyzed samples by analyzing the weathering, erosion, and morphologic analysis of the Martian landscape. MAHLI is a camera suite for the analysis of closeup images of the samples to establish the mineral, textural, and structural contexts.

4.3.2.2 Spectrometers

Several analytical instruments are included in the payload of the rover for the determination and quantification of the chemical composition of the analyzed rocks and regolith: the Alpha-Particle X-Ray Spectrometer (APXS); the Chemistry and Camera (ChemCam), a LIBS spectrometer; Chemical and Mineralogy (CheMin), an X-Ray Diffraction/Fluorescence (XRD/XRF) instrument; or Sample Analysis at Mars (SAM), a suite of three instruments including gas chromatography, mass spectroscopy, and laser spectroscopy, aimed at the detection of elements associated with the potential existence of life.

4.3.2.3 Radiation detectors

Several instruments are onboard the rover dedicated to the following: the characterization of high-energy particles on the surface with the Radiation Assessment Detector (RAD), critical to determine the risks for a potential manned mission; and the detection of subsurface water molecules by the Dynamic Albedo of Neutrons (DAN) instrument, which has astrobiological implications, but also serves to study the potential use of *in situ* water by future missions.

4.3.2.4 Environmental sensors

The meteorological station onboard the MSL is called Rover Environmental Monitoring Station (REMS) monitors temperature, pressure, winds, UV radiation, humidity, etc. The data gathered by REMS are used to characterize and model the Martian climate along the seasons and years, which are key not only to understand the Martian weather, but also for the planning of future manned missions.

4.3.3 Where has curiosity led Curiosity? The science discoveries

The Curiosity rover has been working nonstop since its arrival to Mars in 2012, and it will continue to do so for the time being, as it is still in good shape. Probably the Mars 2020 rover, Perseverance, will arrive to Mars in 2021, while Curiosity is still fully operative. Given the scientific feedback already provided by this rover, it can be considered as a new great success by NASA, but it will for sure still bring interesting new discoveries during the remainder of the mission. Some of the MSL findings have confirmed or supported the knowledge gained by MER, but others have pushed the Martian understanding some steps further.

4.3.3.1 The water: sustained water currents, and fresh water and thick atmosphere

The Curiosity rover identified boulders that likely were rounded by the effect of water currents in Mount Sharp, in what was probably a river/lake system where water flowed for around one million years [35]. Also, the SAM instrument isotopic analysis on the Martian atmosphere elements indicates that the planet has come to be deprived of its early thicker atmosphere and water masses by, among other things, the effects of the solar winds in a planet without a magnetosphere [36–38].

4.3.3.2 A habitable environment

The analysis of the Martian chemistry of mudstones in the Yellowknife Bay confirmed the presence of key elements needed for life, such as oxygen, phosphorus, sulfur, and nitrogen. Also, the presence of fresh water can be inferred from the lack of many salts and the presence of clay minerals [39].

4.3.3.3 Organic carbon

One of the most important discoveries performed by the SAM instrument is based on the analyses from drilled samples (at some centimeters depth) in Mount Sharp. These analyses confirmed the presence of organic molecules. This is a very important discovery considering that any form of life would be formed from organic compounds. Also, it shows that preservation (and detection) of these molecules is possible, even at a few centimeters below the surface (where the UV-radiation degrades and breaks the long molecular chains of the organics) [40].

4.3.3.4 The methane cycle on Mars

The presence of methane on Mars is puzzling the scientific community, as it varies in concentration with time, meaning that there is an existing methane cycle on the planet. The formation of methane can occur from chemical reactions, but also by living organisms. Curiosity has monitored the Martian methane with SAM's Tunable Laser Spectrometer, observing variations up to one order of magnitude in a period of only 2 months. Its source, however, is still to be identified [41].

4.3.3.5 Radiation

The RAD instrument has performed analysis during the whole duration of the mission, including cruise. The results show that the radiation dose during a Mars mission would pose a risk for the human crew. The Galactic Cosmic Rays and the Solar Energetic Particles are the main radiation sources that will affect potential future astronauts on the surface of Mars. The radiation characterization performed by Curiosity during the mission will help defining safe mission concepts for the manned exploration of Mars [42–43].

The Martian missions after MSL have been orbiters centered in the study of the Martian atmosphere. Mars Atmosphere and Volatile Evolution (MAVEN) and the Indian Mars Orbiter Mission (Mangalyaan) were launched in 2013 and have been studying the evolution of the higher layers of the atmosphere in order to understand its loss. MAVEN has confirmed how the solar wind, in absence of a protective magnetic field, facilitates the escape of the charged particles on the Martian atmosphere [44]. Later in 2016, the first part of the ExoMars mission deployed the Trace Gas Orbiter (TGO) in the Martian orbit, incorporating a suite of instruments (ACS and NOMAD) to analyze the concentration of methane and other gases with detection limits as low as 10 parts per thousand, with the ultimate objective of helping understand the methane cycle in Mars.

Finally, 2018 saw the launch of another NASA lander, using the same platform concept as for the Phoenix mission some years before. This mission was named Interior Exploration using Seismic Investigation, Geodesy and Heat Transport (InSight) and was devoted to the analysis of the planet interior, performing seismologic and in-depth thermal analysis. The data gathered by this mission will help understand the formation process of the planet compared to others in the Solar System.

5. What's next? The future of Martian exploration

The decade of the 2020s opens a *golden age* of the *in situ* Martian exploration. During the summer of 2020, two missions of great impact will be launched to Mars. On the one hand, the improved (1 Ton heavy) version of MSL, the Perseverance rover, will be deployed by NASA on the Jezero crater in Mars. On the other hand, China will launch the ambitious Tianwen-1, which will try to place, in only one mission, an orbiter, a lander and a mid-size rover (240 kg). In the following launch window in 2022, Europe (ESA) and Russia (Roscosmos) will join the exploration of Mars with the second phase of the ExoMars mission, which will deploy the Rosalind Franklin rover (named in honor of the British scientific) on Oxia Planum by the beginning of 2023.

5.1 Flying high: the perseverance breakthroughs

The upcoming NASA mission to Mars with the Perseverance rover breaks frontiers in many scientific and technological aspects, also giving clear steps as defined by the Mars Exploration Program. Perseverance is in many aspects similar to Curiosity, but it implements several improvements and novel analytical techniques. On the one hand, Raman spectroscopy, a powerful analytical technique for molecular identification of samples, unprecedented in planetary exploration missions, appears in the payload of the rover not one, but twice. The SuperCam multi-analytical instrument suite includes a remote Raman spectrometer that will analyze rocks and soils at distances of up to 12 m. Also, the SHERLOC instrument placed in the arm of the rover will use an UV laser source to perform Raman spectroscopy optimized for the detection of organics.

Paving the way for future human missions, Perseverance is equipped with the Mars Oxygen ISRU Experiment (MOXIE), a technological demonstrator bound to evaluate the feasibility of extracting pure oxygen out of the CO_2 present in the Martian atmosphere. If working, this technology could be escalated to obtain propellent for vehicles leaving Mars for return to Earth in sample return missions (or to be used during manned stays in the planet in the long term).

Another technological demonstrator is related to the 3-D exploration of Mars, as the Perseverance rover will deploy a helicopter capable of flying in the thin Martian atmosphere. This will demonstrate the feasibility of future drone-based exploration missions on Mars.

Last but not least, a very critical payload in the Martian exploration roadmap is the Sample Caching System. This is one of the most complex robotic systems ever built on a rover and will be used to cache samples deemed interesting by the analytical instruments of the rover, sealing them in tubes that will be left on the Martian surface to be picked up by a rover on a future mission to be returned to Earth. With all these novelties, Perseverance will be setting new milestones in the Mars Exploration Program pathway.

5.2 Drilling down: the collaborative robotic biologist

The ExoMars 2022 mission is the result of a collaborative effort from several points of view. On the one hand, the European Rosalind Franklin rover will be landed by the Russian Kazachok lander, constituting a joint effort and tight collaboration between ESA (and the different European participant countries) and Roscosmos. On the other hand, the rover itself is equipped with a sample preparation and distribution system that has been designed in order to allow a very tight collaboration of all the payload instruments, with a clear objective in mind: to look for traces of life.

The most important novel technology used by the ExoMars rover is a drill that will be used to obtain samples from down to 2 m depth. This element is critical for the mission: this robotic biologist will be able to analyze samples obtained from a depth at which the organic molecules will be much better preserved from the radiation on the surface.

Other interesting features of this rover are the automatic navigation system that will allow an unattended daily navigation of ~100 m, including a novel *walking wheel* design that will be used to avoid getting stuck in sandy terrains.

The rover payload includes a Panoramic Camera (PanCam) and the Infrared Spectrometer for ExoMars (ISEM) placed on the rover mast. These instruments will help mapping the terrain and the sample selection for other instruments. This selection will also be narrowed with the help of the Water Ice and Subsurface Deposit Observation on Mars (WISDOM), a subsurface radar to study the soil stratigraphy under the rover; also the Autonomous Detector of Radiation of Neutrons Onboard Rover at Mars (ADRON-RM) will look for water or hydrated minerals under the Martian surface. Considering the information obtained from all these instruments, a decision will be taken on the drilling site to optimize the chances of detecting biosignatures. There will be several opportunities for analysis as the rover is designed to support several drilling cycles.

The sample analysis will be performed by several instruments in a choreographed sequence starting during the drilling with the Mars Multispectral Imager for Subsurface Studies (Ma_Miss), an IR spectrometer placed on the drill tip. The results from this instrument will be key to understand the rock formation and sedimentation processes without considering the atmospheric influence or weathering, while also helping contextualize the sample texture and structure. The sample will then be extracted and color-imaged by the Close-Up Imager (CLUPI) for morphologic, structural, and textural analysis. Finally, the sample is introduced in the rover body for analysis by the Analytical Laboratory Drawer (ALD), inside an Ultra Clean Zone (UCZ), an area with the highest cleanliness and sterilization requirements to avoid controversies regarding the results obtained by the very sensitive instrumentation of the ALD.

The Rosalind Franklin rover includes a sophisticated sample preparation and distribution system that will crush, dose, and flatten the sample on a carrousel that will move in sequence to allow the analysis of the sample by the three ALD instruments: MicrOmega is an IR spectrometer that will identify the potential regions of interest for analysis by the other instruments. The Raman Laser Spectrometer (RLS), the first ever Raman spectrometer qualified for a space mission, will perform a molecular identification of the materials on the same spots indicated by MicrOmega as regions of interest, and others randomly on the sample surface. And finally, the Mars Organic Molecule Analyzer (MOMA), a gas chromatography/mass spectrometry (GC/MS) instrument that can also work in Laser Desorption/Mass Spectrometer (LDMS) mode, will analyze the sample. With MOMA-LDMS, it will be possible to analyze the very same spots analyzed by RLS and MicrOmega. If the sample is considered interesting, then MOMA-GC/MS can be commanded on the sample (by dosing sample on one-use pyrolysis ovens) to characterize, with very low limits of detection, the organic compounds present on the sample.

This is how the ExoMars rover will perform a sequential collaborative analysis in which all the elements need to work as expected to ensure a successful measurement. This risky but an ambitious approach is necessary to maximize the chances of obtaining a major breakthrough in the exploration of Mars: the possibility of detecting preserved complex organic molecules or biomarkers on the red planet.

When considering all the rover missions to Mars, it can be observed how payloads and mission designs have evolved to give answers to the scientific questions that arise after every new discovery; of course, to the extent that technological advances have allowed. **Figure 2** includes a summary of the payloads included in the Martian rovers.

ROVERS' PAYLOADS DEFINITION

SCIENCE LAYER / ROVER	Sojourner	Spirit & Opportunity	Curiosity	Perseverance	Rosalind Franklin
Atmosphere	IPM (*)	PanCam	MARDI (**) MEDLI (**) MASTCAM	SUPERCAM Mastcam-Z	PanCam
Environment	ASI/MET (*) MPI (*)	Magnet Array	REMS RAD	MEDA	
Surface	IPM (*) RIC APXS	APXS MI Mini-TES MS	CHEMCAM CHEMIN MAHLI APXS SAM MASTCAM	PIXL SUPERCAM SHERLOC WATSON	ISEM PanCam CLUPI
Sub-surface			DAN	RIMFAX	WISDOM Ma_MISS MicrOmega ADRON RLS CLUPI MOMA
Engineering	WAE Wheel abrasion MAE (dust on solar panels) RIC Wheels Camara	RAT	NavCam HazCam	NavCam HazCam MOXIE	

Cameras Sensors Spectrometers Detectors

(*) on Pathfinder Lander
(**) during descent

Figure 2.
Martian rovers' payloads and analysis objectives on Mars.

5.3 Bring it home: the Mars sample return

Converting the planetary exploration paradigm into a world effort instead of national initiatives will be necessary for the mid-term plans of Mars exploration and will require a tight collaboration between agencies in order to make substantial advances in the coming decades. As soon as technology readiness allows it, returning samples to Earth for analysis with the best available instrumentation is the next reachable step in the exploration effort. In this framework, the Mars Sample Return (MSR) program is designed as a joint effort between NASA and ESA, defining a complex sequence of missions beginning with the Perseverance rover, selecting and caching the first samples. Later, an ESA rover will be sent sometime during the decade (2026?) to retrieve and store the samples to be placed in a Mars Ascent Vehicle (MAV) that will place the samples in orbit, where they will wait until a return-trip ship captures them to bring them to Earth for analysis.

These complex mission designs, launches, and operations' sequence will constitute a major milestone in the Martian exploration, where the next step will be the design of *in situ* manned missions to Mars.

6. Conclusions

The robotic exploration of Mars is a consequence of humanity's awe toward our red neighbor and has become a reality when technology has reached the needed maturity, also influenced by other socio-economical aspects. The space race of the 1960s and the 1970s was the starting point of the Martian exploration and is a good example of how technological development was pushed beyond unimaginable limits thanks to the social, political, and economic support. In the late 1970s, however, the race had already been won over by the United States, and economies were suffering the petrol crisis. This resulted in a loss of momentum in the exploration missions to Mars, and the missions to Mars stopped until the implementation of the Martian Exploration Program in the 1990s.

The MEP is an ambitious program for the exploration of Mars, which was conceived as an exploration effort based on an international collaboration, aimed at joining efforts among the different space agencies worldwide. This set a new era in the exploration of Mars, where not only collaboration between Agencies is needed for mission's preparation, but also cooperation is required among different instruments once on Mars to give proper answers to scientific challenges. This has resulted into complex rover designs and missions where a suite of experiments works altogether for a common goal.

The decade of the 2020s will set new milestones in the Martian exploration paradigm, not only for pushing the technological limits and conquering new dimensions of explorations (helicopters, subsurface drilling), but also achieving the final step in the Mars studies evolution observations (with telescopes from Earth, from orbit, *in situ* from surface, etc.), bringing home (Earth) Martian samples for its study, paving the way for the human exploration of the red planet.

In this new era, the collaboration between instruments, rovers, missions, and, finally, agencies and political actors will be key to obtain the best results to, ultimately, unravel the mysteries of our red neighbor and, who knows, maybe answer one of the most transcendental questions of humankind: Are we alone?

Acknowledgements

Authors acknowledge Spanish institutions: Instituto Nacional de Técnica Aeroespacial (INTA) and Universidad de Valladolid (UVa), as supporting institutions for authors professional activities, and Ministerio de Ciencia e Innovación (MICINN) for its funding support through grants ESP2014-56138-C3-1-R, ESP2014-56138-C3-2-R, ESP2107-87690-C3-1-R, ESP2107-87690-C3-3-R.

Conflict of interest

The authors declare that they have no conflicts of interest with respect to research, authorship and/or publication of this book chapter.

Author details

Andoni G. Moral Inza[1]* and Guillermo Lopez-Reyes[2]

1 Instituto Nacional de Técnica Aeroespacial (INTA), Torrejón de Ardoz, Spain

2 Unidad Asociada UVa-CSIC-CAB, Universidad de Valladolid, Valladolid, Spain

*Address all correspondence to: moralia@inta.es

References

[1] Anderson HR. Mariner 4 measurements near Mars, initial results, spacecraft description and encounter sequence. Science. September 1965;**149**(3689):1226-1228. DOI: 10.1126/science.149.3689.1226

[2] Leighton, Robert B, Murray BC, Sharp RP, Allen JD, Sloan RK. Mariner IV photography of Mars: Initial results. Science, New Series. 1965;**149**(3684):627-630. DOI: 10.1126/science.149.3684.627

[3] Smith EJ, Davis L Jr, Coleman PJ Jr, Jones DE. Magnetic field measurements near Mars. Science, New Series. 1965;**149**(3689):1241-1242. DOI: 10.1126/science.149.3689.1241

[4] Van Allen JA, Frank LA, Krimigis SM, Hills HK. Absence of Martian radiation belts and implications thereof. Science, New Series. 1965;**149**(3689):1228-1233. DOI: 10.1126/science.149.3689.1228

[5] Leighton RB et al. Mariner 6 and 7 television pictures, preliminary analysis. Science. 1969;**166**:49-67. DOI: 10.1126/science.166.3901.49

[6] Herr KC, Pimentel GC. Evidence for solid carbon dioxide in the upper atmosphere of Mars. Science. 1970;**167**:47-49. DOI: 10.1126/science.167.3914.47

[7] Stillman CC. Infrared radiometer for the 1969 Mariner mission to Mars. Applied Optics. 1969;**8**:639-643. DOI: 10.1364/AO.8.000639

[8] Hartmann WK. The New Mars: The Discoveries of Mariner 9. Prepared for the NASA Office of Space Science /William K. Hartmann and Odell Raper. LC Classification QB641. H33; 1974. Available from: https://lccn.loc.gov/74600084

[9] Farmer CB et al. Viking—Mars atmospheric water vapor mapping experiment—Preliminary report of results. Science. 1976;**193**:776-780. DOI: 10.1126/science.193.4255.776

[10] Jakosky BM, Farmer CB. The seasonal and global behavior of water vapor in the Mars atmosphere: Complete global results of the Viking atmospheric water detection experiment. Journal of Geophysical Research. 1982;**87**(B4):2999-3019. DOI: 10.1029/JB087iB04p02999

[11] Kieffer HH et al. Infrared thermal mapping of the Martian surface and atmosphere—First results. Science. 1976;**193**:780-786. DOI: 10.1126/science.193.4255.780

[12] Blasius KR, Cutts JA, Guest JE, Masursky H. Geology of the Valles Marineris: First analysis of imaging from the Viking 1 Orbiter primary mission. Journal of Geophysical Research. 1977;**82**(28):4067-4091. DOI: 10.1029/JS082i028p04067

[13] Hargraves RB et al. Viking magnetic properties experiment: Extended mission results. Journal of Geophysical Research. 1979;**84**(B14):8379-8384. DOI: 10.1029/JB084iB14p08379

[14] Clark BC et al. Inorganic analyses of Martian surface samples at the Viking landing sites. Science. 1976;**194**(4271):1283-1288. DOI: 10.1126/science.194.4271.1283

[15] Hess SL et al. Meteorological results from the surface of Mars: Viking 1 and 2. Journal of Geophysical Research. 1977;**82**(28):4559-4574. DOI: 10.1029/JS082i028p04559

[16] Lazarewicz AR et al. The Viking Seismometry Final Report. NASA CR-3408; 1981. p. 59

[17] Levin GV, Straat PA. Recent results from the Viking labeled release experiment on Mars. Journal of Geophysical Research. 1977;**82**(28):4663-4667. DOI: 10.1029/JS082i028p04663

[18] Shirley D, McCleese DJ. Mars Exploration Program Strategy: 1995-2020. American Institute of Aeronautics and Astronautics 96-0333; 1996. DOI: 10.2514/6.1996-333

[19] Bolden CF Jr. NASA Strategic Plan 2014. NP-2014-01-964-HQ [Internet]. 2014. Available from: https://www.nasa.gov/sites/default/files/files/FY2014_NASA_SP_508c.pdf [Accessed: 26 June 2020]

[20] Edwards CD Jr et al. Relay Orbiters for Enhancing and Enabling Mars In Situ Exploration. NASA Science Solar System Exploration. Planetary Science Decadal Survey; 2013-2022. Available from: https://solarsystem.nasa.gov/studies/54/relay-orbiters-for-enhancing-and-enabling-mars-in-situ-exploration/ [Accessed: 26 June 2020]

[21] Schofield JT et al. The Mars pathfinder atmospheric structure investigation/meteorology (ASI/MET) experiment. Science. 1997;**278**(5344):1752-1758. DOI: 10.1126/science.278.5344.1752

[22] Hviid SF et al. Magnetic properties experiments on the Mars Pathfinder lander: Preliminary results. Science. 1997;**278**(5344):1768-1770. DOI: 10.1126/science.278.5344.1768

[23] Smith PH et al. Results from the Mars Pathfinder camera. Science. 1997;**278**(5344):1758-1765. DOI: 10.1126/science.278.5344.1758

[24] Rieder R et al. The chemical composition of Martian soil and rocks returned by the mobile alpha proton X-ray spectrometer: Preliminary results from the X-ray mode. Science. 1997;**278**(5344):1771-1774. DOI: 10.1126/science.278.5344.1771

[25] Foley CN et al. Final chemical results from the Mars Pathfinder alpha proton X-ray spectrometer. Journal of Geophysical Research. 2003;**108**(E12):8096. DOI: 10.1029/2002JE002019

[26] Richard AK. Opportunity tells a salty tale. Science. 2004;**303**(5666):1957. DOI: 10.1126/science.303.5666.1957b

[27] Squyres SW et al. Ancient impact and aqueous processes at Endeavour crater. Mars Science. 2012;**336**(6081):570-576. DOI: 10.1126/science.1220476

[28] Morris RV et al. Identification of carbonate-rich outcrops on Mars by the Spirit rover. Science. 2010;**329**(5990):421-424. DOI: 10.1126/science.1189667

[29] Squyres SW et al. Detection of silica-rich deposits on Mars. Science. 2008;**320**(5879):1063-1067. DOI: 10.1126/science.1155429

[30] Squyres SW et al. Pyroclastic activity at home plate in Gusev crater. Mars Science. 2007;**316**(5825):738-742. DOI: 10.1126/science.1139045

[31] Squyres SW et al. Overview of the opportunity Mars exploration rover mission to Meridiani Planum: Eagle crater to purgatory ripple. Journal of Geophysical Research. 2006;**111**(E12S12). DOI: 10.1029/2006JE002771

[32] Landis GA et al. Dust devils in gusev crater: A second year of observations by the spirit rover. Seventh International Conference on Mars. 2007. Available from: https://www.lpi.usra.edu/meetings/7thmars2007/pdf/3149.pdf

[33] Smith M et al. First atmospheric science results from the Mars exploration rovers miniTES. Science. 2004;**306**(5702):1750-1753. DOI: 10.1126/science.1104257

[34] Schröder C et al. Meteorites on Mars observed with the Mars exploration rovers. Journal of Geophysical Research, Planets. 2008;**113**(E6):E06S22. DOI: 10.1029/2007JE002990

[35] Williams RME et al. Martian fluvial conglomerates at Gale crater. Science. 2013;**340**(6136):1068-1072. DOI: 10.1126/science.1237317

[36] Mahaffy PR et al. Abundance and isotopic composition of gases in the Martian atmosphere from the Curiosity rover. Science. 2013;**341**(6143):263-266. DOI: 10.1126/science.1237966

[37] Mahaffy PR et al. The imprint of atmospheric evolution in the D/H of Hesperian clay minerals on Mars. Science. 2015;**347**(6220):412-414. DOI: 10.1126/science.1260291

[38] Webster CR et al. Isotope ratios of H, C, and O in CO_2 and H_2O of the Martian atmosphere. Science. 2013;**341**(6143):260-263. DOI: 10.1126/science.1237961

[39] Grotzinger JP et al. Habitability, taphonomy, and the search for organic carbon on Mars. Science. 2014;**343**(6160):386-387. DOI: 10.1126/science.1249944

[40] Freissinet C et al. Organic molecules in the Sheepbed Mudstone, Gale crater, Mars. Journal of Geophysical Research. 2015;**120**(3):495-514. DOI: 10.1002/2014JE004737

[41] Webster CR et al. Mars methane detection and variability at Gale crater. Science. 2015;**347**(6220):415-417. DOI: 10.1126/science.1261713

[42] Zeitlin C et al. Measurements of energetic particle radiation in transit to Mars on the Mars Science Laboratory. Science. 2013;**340**(6136):1080-1084. DOI: 10.1126/science.1235989

[43] Hassler DM et al. Mars' surface radiation environment measured with the Mars Science Laboratory's Curiosity rover. Science. 2014;**343**(6169):1244797. DOI: 10.1126/science.1244797

[44] Ramstad R, Brain DA, Dong Y, et al. The global current systems of the Martian induced magnetosphere. Nature Astronomy. 2020. DOI: 10.1038/s41550-020-1099-y

Section 3

Human - Rated Missions

Chapter 7

Human Missions Analysis for Intelligent Missions Improvement

Carole Tafforin

Abstract

The topic of this chapter is not manned vs. robotic missions but how to integrate them for successful missions. The point of view is from that of a human observer, an ethologist, with the goal to gain further knowledge on human behavior and high technology readiness levels in the field of exploration missions. On one hand, the concept is the adaptability of men/women to be trained and on the other hand, it is the reliability of artificial intelligence systems to be incremented. The content of the chapter is: (i) ethological analysis based on numerical methods; (ii) strategy, cooperation, and adaptation; (iii) artificial intelligence and emotional intelligence; and (iv) man-rated Mars exploration demands.

Keywords: methods, human factors, robotics, artificial intelligence, earth simulation, analog environments

1. Introduction

The challenge of the astronaut Jean-Loup Chrétien's Extra-Vehicular Activity (EVA) during the Aragatz orbital flight was to expand a deployable structure following operational procedures. Many hours of training were performed in a swimming pool to simulate each technical sequence and to test out the material in an optimal way. The operation did not run in orbit and the operator kicked in the structure to be deployed. We will not debate on manned vs. robotic missions but we will open discussions on how to integrate them for successful missions. The point of view is from a human observer, an ethologist, with the goal to gain further knowledge on human being and high technology readiness levels in the field of exploration missions.

From an evolutionary perspective, the naturalist Charles Darwin developed the idea that behavior is an important element of competition and natural selection in "The Origin of Species" [1]. His trip aboard the Beagle around the world for observing animals in their natural living conditions and the related geological environments is a fundamental event in human history. Over a long-term dynamics process, the motor behavior has progressed step by step in motion and motor patterns adapted to terrestrial gravity. Then, mankind developed elaborate technologies to fly further and further away and discovered how to move under reduced gravity on Moon and in low-earth orbits. EVA gave man the ability to indifferently operate in a three-dimensional space with head-up and head-down. Orientations, postures, and movements have to be coordinated under weightlessness according to new processes [2, 3]. Sensory-motor functions and cognitive functions are deeply demanded. Other physiological and psychological functions are necessary and become more

Figure 1.
Evolutionary steps from Earth to Space (extract from picture © Cité de l'Espace—Toulouse).

prominent over time. Thus, the next big step of human beings from earth to space (**Figure 1**) is to link up human individual intelligence with artificial intelligence (AI) for intelligent missions improvement.

2. Ethological analysis based on numerical methods

Since 30 years in the space domain [4, 5], ethology, defined as the science of behavior, has covered a wide panel of application fields on men and women living and working in real situations (orbital flights, parabolic flights), in simulated conditions (isolation and confinement campaigns), and in analog environments (south-polar stations, north-pole expeditions).

In its operational definition "behavior is the expression, the emergence, here and now, of a historical system constituted by the individual and his own universe" [6].

Behavior falls within the scope of phenomena and not physico-chemical objects. This phenomenon appears as a motor activity performed by a subject and represents a continuous manifestation. There is always an action or fact to describe even when the subject is at rest. Such a description leads to a large volume of discursive

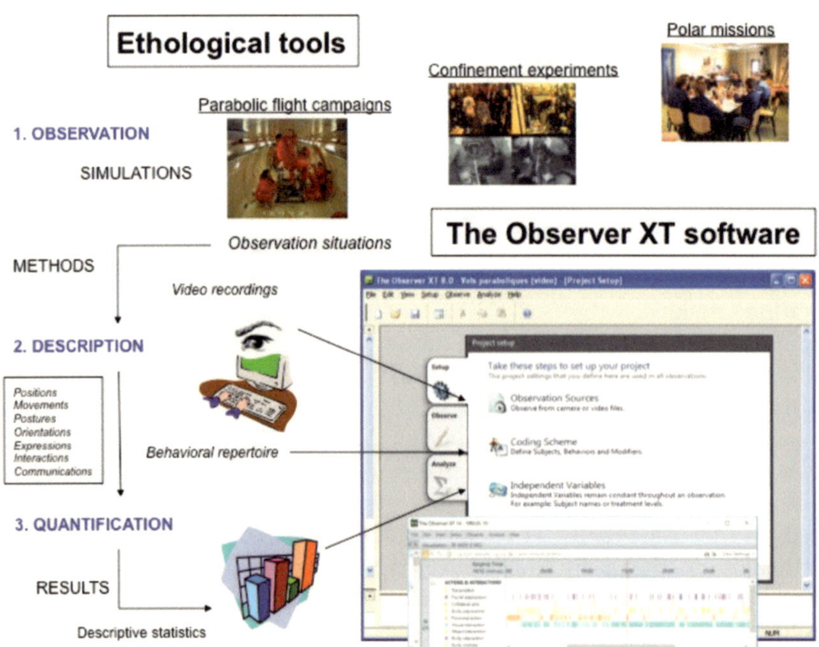

Figure 2.
Ethological tools related to The Observer XT software.

data whose subsequent analysis must be sequenced. The specificity of the ethological method is the transition from qualitative observation to quantitative description that reflects continuity in its analogical aspects as a mathematical function. The analysis is carried out in three steps: observation, description, and quantification (**Figure 2**).

2.1 Observation

In human ethology, the method is to observe the subject's behavior in usual activities. It can be not only acts of daily life, leisure, execution of work tasks but also in experimental situations or the performance of tests. A special feature of this approach is to exploit the field of observable events. Everything is observable to the extent that the relationship between the individual and his environment emerges, that is, is objectively visible. The observer's eye remains irreplaceable for describing behavior in all its complexity (positions, orientations, motions, expressions, interactions, and communications).

2.2 Description

Since behavior is a continuous process, it is necessary to build a system of units that breaks down this continuity and digitizes it. This amounts to dividing the observed behavioral flow into motor acts, the various possibilities of which are described over time.

The behavioral reading instrument consists of establishing a repertoire, in terms of action verbs, according to a vocal-mimo-posturo-gestural vocabulary (e.g., "raise the arm," "lower the head," "turn the body to the right," "darken the eyebrows," "smile," "speak to," "give an object to," etc.). The ethologist builds a digitization instrument corresponding to different levels of description (microscopic vision, macroscopic vision).

The description tool is the direct input in the field or the delayed input of video recordings collected in a situation.

2.3 Quantification

The numerical tool measures the probability of occurrence of each act in the repertoire, by counting these manifestations in terms of frequency, duration, or sequential order, replacing each behavioral unit in its functional framework and in its own space. It allows this complexity to be represented by multivariable processing. The quantification of the observed behavior is then translated into occurrence frequencies (absolute or relative), transition frequencies, and association frequencies. The duration of the items is also measured in the behavioral sequence. A spatial mapping into digitized units is performed in the same way. We obtain a scheme of use of space combining space, time, and activity.

2.4 Computer support

A software-based solution for research in space ethology is called The Observer XT® software [7]. It is a professional system that can be used for the collection, analysis, and presentation of observational data. It allows annotation of behavioral descriptors through a traditional encoding process. Its technical specificity is the synchronization of video files with the collection of ethological data as state events or point events, and any other source of information such as psycho-physiological measurements and environmental parameters.

Right now, the interface between the human observer and *The Observer XT®* software is not computerized. Because of the complexity of the behavior to be analyzed, the ethologist's eye remains an essential tool. We can use techniques like the newly developed Facial Action Coding System [8] first adopted by psychologists [9]. It is a common standard to systematically categorize the physical expression of emotions, but it does not integrate the behavioral activity from the human repertoire as a whole, that is, in egocentric references (with regard to the subject), allocentric references (with regard to the other subjects), and geocentric references (with regard to the three-dimensional environment).

3. Strategy, adaptation, and cooperation

Such an approach is not only concerned with the result of the behavior, that is, performances, but also with the motor patterns leading to it, that is, behavioral strategies. They are a synergy of abilities that could be a high-level plan to achieve goals under extreme conditions involving both strategic planning and strategic thinking [10]. As a cognitive activity, it produces decision-making and thus a motor activity follows to perform the task for the success of the mission goal.

During the Aragatz mission, the astronaut in EVA carried out procedures according to a checklist consistent with the operation called ERA. The final technical condition was a deployed structure (**Figure 3**). The final human action was a kick in the compact package (**Figure 4**).

During simulations in water immersion, the operator followed motor sequences as we analyzed them with numerical methods. Strain arms followed by grasp hands followed by release hands followed by keep still, then manipulate followed by release hand followed by flex arms, are motor patterns in neutral buoyancy. The task automation

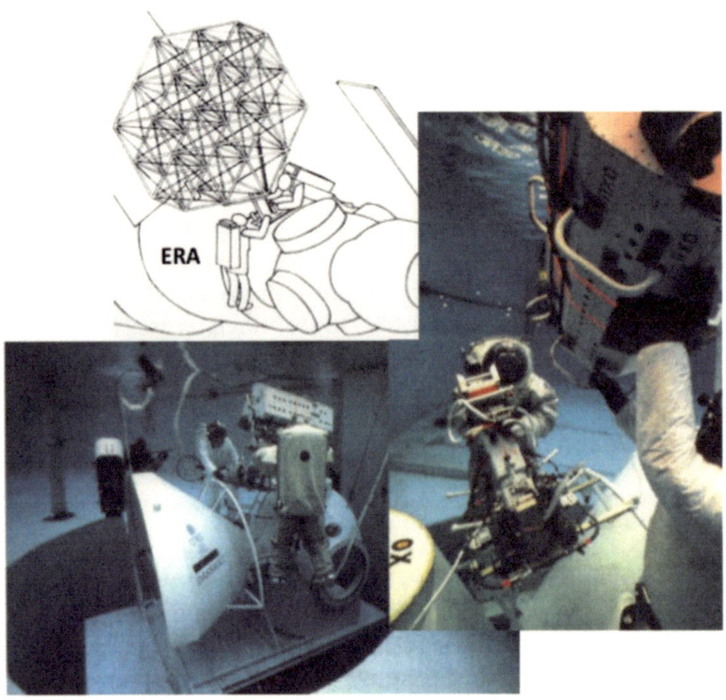

Figure 3.
EVA in water immersion during the ERA operation.

Contingency Table Analysis

DF:	15
Total Chi-square:	474.522 p< 0.0001
G Statistic:	.
Contingency Coefficient	.443
Cramer's V:	.495

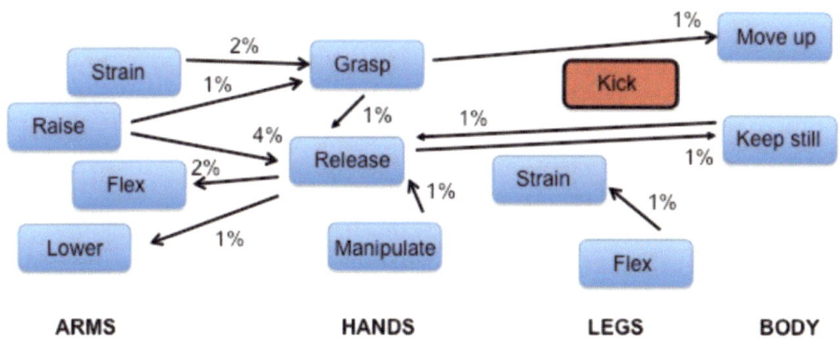

Figure 4.
Motor sequences of the astronaut during the ERA operation in orbital flight (red) and in swimming pool (blue).

could be performed by a robot as by the astronaut in nominal situation. Because of off-nominal conditions with the negative goal of failure, Jean-Loup Chretien's strategy was to add a new action that was not learned and not repeated before but efficient for a positive final goal. On one hand, there is the reliability of robots and on the other hand, there is the adaptability of man to his environmental conditions. In ethology, the behavioral manifestations are optimized relationships between the individual and the environment. Such adaptive strategies can be built as an action pattern occurs in an environment [11] with unexpected events or untrained conditions.

Adaptation capacities could be factors of human intelligence. In psychology, the intelligence quotient is a total score derived from standardized tests obtained by dividing a person's mental age score [12] but not necessary from measurements of coping skills. In biology, humans face basically the same adaptive challenges as all organisms on earth but they are complex in having most of their adaptation transmitted culturally upon experience.

Quantitative descriptions of the sensory-motor adaptation from the early seconds in parabolic flights to the first days and last days in orbital flights, then on the social adaptation over long-duration analog missions for future Mars exploration, offer an overview of human intelligence.

We found spontaneous [13], preliminary, and integrative stages of behavioral adaptation, emphasizing new relationships between the body references and those of the surrounding world [14]. Such experiences lead the subject to develop a new mental representation of space [15]. The adaptive model refers to the sensory-motor sphere and neuro-physiological sphere like a "hard" system that acts to recover its main equilibrium in terms of conservative regulation with respect to a mismatching physical environment. At the highest levels, human brain is one of the most amazing systems as biological organ, functional machine or supercomputer [16]. Connections are done through a "soft" system with new adaptive strategies. For instance, stage by stage, the astronaut's motor actions are to manipulate floating objects and to move upside down, which show he is exploiting new possibilities of the weightlessness conditions. Human intelligence is incremented with motor experience just as advanced machine learning.

In this regard, the issues currently being raised are on one side, to what extend human adaptability is required for Mars exploration and on the other side, to what extend machine learning capability is involved for Mars exploration?

Cooperation should be emphasized. We need to take carefully into account the Human Factors (HFs) in regard to their diversity and the quality of relations between heterogeneous partners: human-human; man-robot or AI; machine-machine [17]. When the astronaut Jean-Loup Chretien makes a decision and finds a solution to anomalies in the equipment, thus HF are positive. But operational error detection has to be improved for preventing negative HFs. For instance, automated techniques for routine monitoring during space operations may be appropriate in future missions [18]. Nevertheless, an ethological monitoring performed during routine operations at a Networks Operation Center (NOC) on ground showed increasing human-human verbal interactions as optimizing behavior in the task progress [5]. Cooperative systems could be implemented.

One improvement facilitator is while a crew has to cope with monotony, robot or AI can implement automatic tasks. During long-duration isolation and confinement periods, human behavior is cyclic over time for breaking up monotonous tasks [19] whereas a robot or AI is constant over time and can supplement. Another example, Rover curiosity helps in enriching the curiosity of earthmen and updates their knowledge of planet Mars [20]. It becomes the eyes of human observer in that cooperation.

4. Artificial intelligence (AI) and emotional intelligence (EI)

Intelligent automation and trusted autonomy are being introduced in aerospace cyber-physical systems to support diverse tasks including decision-making, data processing, information sharing, and mission execution with the technological developments of sensor networks [21]. This leads to the field of AI in manned space missions. Men and women are endowed with sensation, feeling, and perception that are altered due to stress, mental and physical workload during living and working simulations of Mars conditions [22]. This leads to the field of EI within the space crewmembers

AI is a virtual concept or rather a set of concepts and technologies [23] more than an independent discipline. It is data integration-dependent in computer sciences vs. a real concept with human decision-making in neurosciences and sciences of behavior. Notions of neural networks are common between human mechanisms and intelligent machines. When the first ones reach limits of physical capacities, the last ones can supplement them to improve long-term operational performance. Significant development in technical innovation has succeeded in transforming manual and repetitive tasks. AI reduces the quantity and improves the quality in many application fields. They could be industrial, intellectual, social [24] and also medical as required by telecommunications between ground control teams and space crews. Security and safety are improved as a result. Benefits of expert systems were aboard the International Space Station (ISS). They make excellent monitoring tools since they never get bored or tired, are always alert, and react faster than astronauts [25]. The status of a trained machine is immediate whereas the human adaptive state is dynamic as it gains in experience during simulations on Earth. For instance, a review of studies about women who lived and worked on remote and isolated Antarctic stations for up to 15 months showed mitigated feelings from positive experiences in the natural physical environment and negative experiences in the social male-dominant context [26]. Such human missions analyses help in drawing scenarios of Mars via Moon missions for intelligent missions (**Figure 5**).

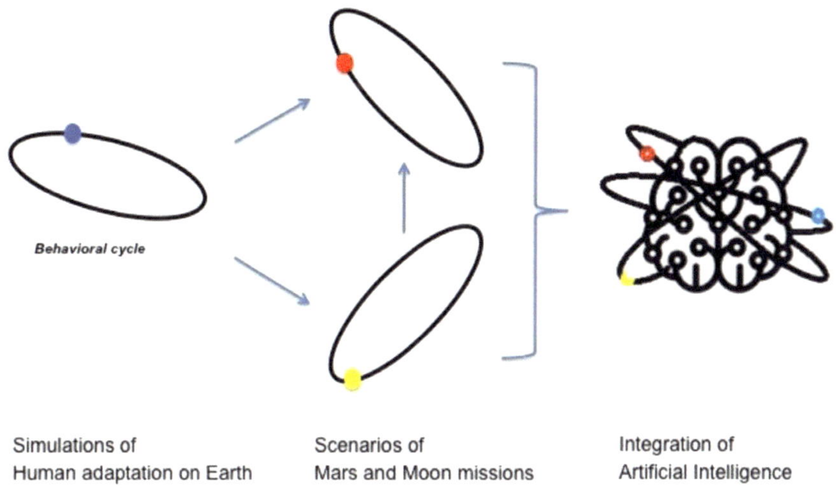

Figure 5.
Artificial intelligence integrated into space missions based on human Earth experiences.

EI has been defined as a wide array of individual variables based on emotional awareness, an ability that has usually been conceptualized along with the cognitive functions [27] and abilities [28]. Personality traits convey the adaptability of intelligence and the subjective experiences based on emotions. Mental health-related outcomes such as well-being or good group spirit are attributed to humans. They impact space missions positively. The key factor is the quality—not necessarily the quantity—of the relationships between the crewmembers and their mental representations within the social context and environmental conditions (**Figure 6**) that activate cognitive functions. AI can benefit from this higher level of integration. Nevertheless, it cannot integrate the uniqueness of an individual as living being. In that process, the human brain works at the highest operational level. In ethology, we also explain behavioral universals by their functions, like smiling observed in

Figure 6.
Cognitive schema of the relationship between crewmembers during interplanetary missions.

autistic persons with communication disorders as well in ethnic groups with their own languages [29]. They can be expressions of emotion and are built by imitation in the human-human relationship. Such associative communication could be a computerized process in human-machine learning and thus create connections between EI and AI. Cultural variable with terrestrial life imprint is another issue to live in space.

5. Man-rated Mars exploration demands

> *"Everything we love about civilization is a product of intelligence, so amplifying our human intelligence with artificial intelligence has the potential of helping civilization flourish like never before – as long as we manage to keep the technology beneficial [30]."*

During a Mars exploration, the crew will be isolated from any civilization. It will be extracted from the ongoing relationships it has experienced on earth, and will be associated to a new micro-society with its own spatial restriction, social deprivation and cultural organization [31]. All of these variables are exacerbated with time. In anthropology, the qualitative description of living rules, working habits, specific customs, and values of remote tribes allow to understand how they behave differently than other self-sufficient groups [32] and how they survive.

Autonomy of the crew [33, 34] is one demand for Mars exploration in technology for life-Support systems [35], and on humanity for a controlled environmental system and an ecological evolution. This advances the field of collective intelligence for survivability from earth to space.

Heterogeneity of the crew [36] is a second demand. Future crews of three to six members, mixed-gender and multi-cultural compositions would be core features of terrestrials gathered for deep space exploration [37]. The operating rules of an isolated and confined crew could thus be compared to the laws governing self-organizing systems. These laws are based on the heterogeneity of their own elements like a thermodynamic model. AI could serve such a group model. It is a dynamic organization where all the forces are in equilibrium and regulated to obtain optimal efficiency structure [38] with multi-function integration [39].

The quantitative description of behavioral expressions of groups who differ according to gender variable, nationality variable, and time variable shows to what extent the value of diversity is a key element along with mission duration (**Figure 7**). We found differences in positive facial expressions ("smile," "laugh") and collateral acts ("scratch the head" "rub the nose") from comparative analysis in three settings during three periods of time. Firstly, the Mars Desert Research Station (MDRS-14d) was a 14-day campaign located in Utah, USA. Secondly, the Controlled Ecological Life Support system experiment (CELSS-180d) was a 180-day confinement that took place in Shenzhen, China. Thirdly, the Mars-500 experiment (MARS-520d) was a 520-day confinement that took place in Moscow, Russia, with the objective to simulate a round trip to Mars.

We observed that the rate/min of facial expressions vs. collateral acts is not significant in the international groups (MDRS-14d and MARS-520d) compared to the mono-national group (CELSS-180d) regardless the simulation time. Well-being expressed by facial expressions highly occurs in the mixed-gender group (MDRS-14d) whereas perceived stress expressed by negative collateral acts occurs in the men's group. This underscores the importance and complexity of personal, social, and cultural factors in behavioral occurrences and the related emotional feelings.

The question then arises about AI as a complex entity like a crew?

DOI: http://dx.doi.org/10.5772/intechopen.90795

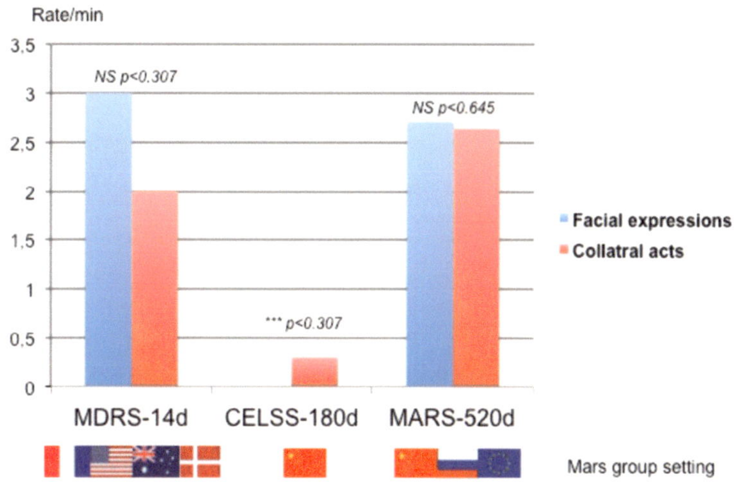

Figure 7.
Behavioral expressions according to three group settings of Mars trip simulations (MDRS-14d: 3Ω & 3ψ; FR,US,AU,DK; CELSS-180d: 5Ω & 1ψ; CH; MARS-520d: 6Ω; RU, EU,CH).

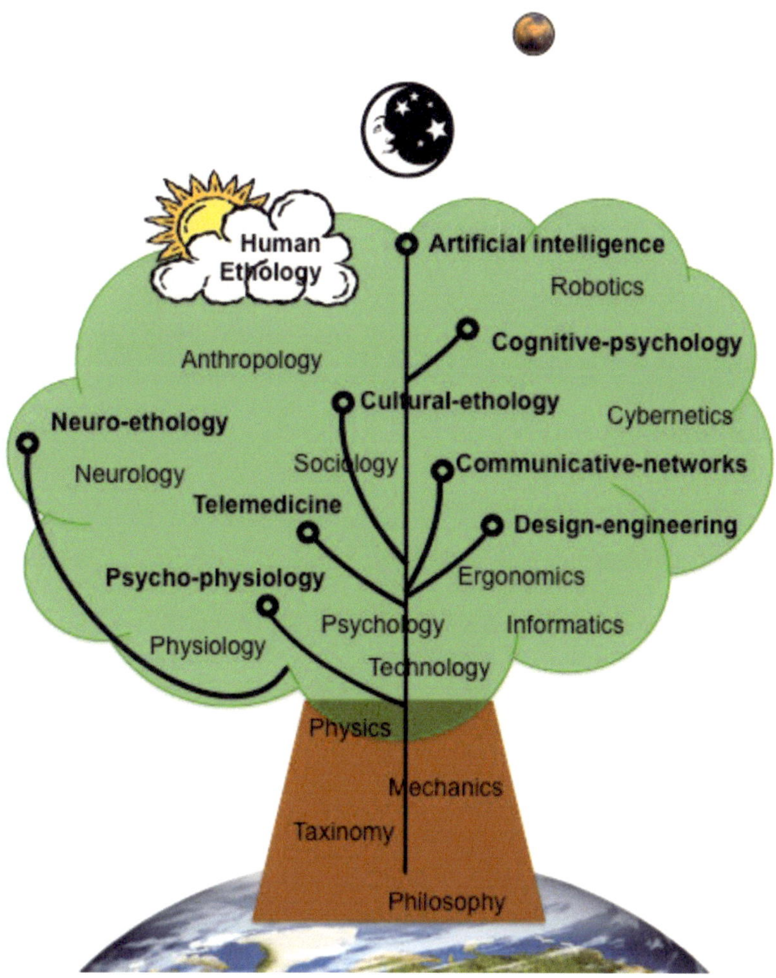

Figure 8.
Epistemological points of multidisciplinary approaches in human ethology.

Man-rated Mars exploration demands are to consider the personal value [40], the mixed-gender [41], and the cultural value [42] of a human mission and to take into account this diversity value as inputs of AI like variations of outputs generated by the machine. An intelligent mission becomes a multi-variable system with strategic processes, adapted processes, and cooperative processes that are running in synergy and that inject multi-sensorial inputs in the crew's EI. We hypothesize that such multiplicity improves over time. Monotony is prevented by robotic-automation tasks and autonomy is ensured by human-dependent actions.

The global demand is multidisciplinary in order to make human-machine interaction evolve. The evolutionary model of the human species is punctuated by so many disciplines that are points of organization from philosophical roots, to the naturalist current, to human ethology (**Figure 8**). Aristotle described behavioral traits from the perspective of a classification, that is, species taxonomy, to complete the anatomical and morphological image. From these epistemological nodes, integrated approaches such as psycho-physiology, neuro-ethology, communicative-networks, telemedicine, cognitive-psychology, and AI as the very last point of integration and evolution of the human species on other planets flourish. The whole approach is systemic [43] and is both behavioral and computational.

6. Conclusion

Numerical methods developed in ethology of humans in space have contributed to give insight into adaptive behaviors that underlie human intelligence as multiple processes of optimization of the individual-environment relationship. Human being's activity is the heart of the system, be it natural or artificial. AI can perform additional functions by processes of imitation, cooperation, and automation that include the diversity of the crewmembers, cyclicity of their behavior, and autonomy of the confined and isolated crew from Earth.

From the individual to the cultural crews' behavior that expresses individual intelligence to collective human intelligence, space missions will improve as intelligent missions by associating AI as follows:

$$\text{Artificial intelligence} + \sum \text{Human intelligences} * = \text{Intelligent missions}$$
$$(* \text{ Individual intelligence} + \text{Emotional intelligence} + \text{Collective intelligence})$$

Acknowledgements

The author thanks the French Space Agency for its scientific and financial support in ethological studies.

Author details

Carole Tafforin
Ethospace, Toulouse, France

*Address all correspondence to: ethospace@orange.fr

References

[1] Darwin C. The Origin of Species. London: John Murray, Albemable Sreet; 1859

[2] Tafforin C. The relationships between orientation, movement and posture in weightlessness: Preliminary ethological observations. Acta Astronautica. 1990;**21**:271-280

[3] Tafforin C. Ethological analysis of human coordinated movements in weightlessness: Comparison between the first and last days of short-term space flights. Journal of Human Movement Study. 1993;**24**:119-133

[4] Tafforin C, Thon B, Guell A, Campan R. Astronaut behavior in orbital flight situation: Preliminary ethological analysis. Aviation, Space, and Environmental Medicine. 1989;**60**:949-956

[5] Tafforin C, Michel S, Galet G. Ethological approach of the human factors from space missions to space operations. In: Pasquier H, Cruzen CA, Schlidhuber M, Lee YH, editors. Space Operations: Inspiring Humankind's Future. Vol. 30. Switzerland: Springer; 2019. pp. 779-794

[6] Campan R. L'animal et son univers. Toulouse: Privat; 1980

[7] Tafforin C, Gerebtzoff D. A software-based solution for research in space ethology. Aviation, Space, and Environmental Medicine. 2010;**81**:951-956

[8] Freitas-Magalhaes A. Facial Action Coding System 3.3: Manual of Science Codification of the Human Face. Porto: FEELab Science Books; 2018

[9] Ekman P, Friesen WV, Hager JC. Facial Action Coding System: A Technique for the Measurement of Facial Movements. Palo Alto: Consulting Psychologists Press; 1978

[10] Mintzberg H, Quinn JB. The Strategy Process: Concepts, Contexts, Cases. Upper Saddle River: Prentice Hall; 1996

[11] Freedman L. Strategy: A History. Oxford: Oxford University Press; 2015

[12] Wechsler D. Wechsler Adult Intelligence Scale - WAIS-R. 1981; Cited in Boake C. From the Binet–Simon to the Wechsler–Bellevue: Tracing the History of Intelligence Testing. Journal of Clinical and Experimental Neuropsychology. 2002;**24**:383-405

[13] Tafforin C. Initial moments of adaptation to microgravity of human orientation behavior, in parabolic flight conditions. Acta Astronautica. 1996;**38**:963-971

[14] Tafforin C. Synthesis of ethological studies on behavioral adaptation of the astronaut to space flight conditions. Acta Astronautica. 1994;**32**:131-142

[15] Tafforin C. Global ethological analysis of crew's adaptive dynamics in extreme settings for future human planet exploration. In: Proceedings of the IAA Symposium on the Future of Space Exploration. Torino, Italy; July 2015

[16] Ornstein R, Thompson RF. The Amazing Brain. New York: Houghton Mifflin Company; 1986

[17] Grès S, Tognini M, Le Cardinal G, Zalila Z, Gueydan G. Cooperation and dialogical modeling for designing a safe human space exploration mission to Mars. In: Proceedings of the IAA Symposium on the Future of space exploration: Towards the stars. Torino, Italy; July 2013

[18] Trollope E, Dyer R, Francisco T, Miller J, Griso MP, Argemandy A. Automated techniques for routine

monitoring and contingency detection of LEO spacecraft operations. In: Pasquier H, Cruzen CA, Schlidhuber M, Lee YH, editors. Space Operations: Inspiring Humankind's Future. Vol. 30. Springer; 2019. pp. 779-794

[19] Tafforin C. The Mars-500 crew in daily life activities: An ethological study. Acta Astronautica. 2013;**91**:69-76

[20] Andrews RG. NASA's curiosity rover finds unexplained oxygen on Mars. In: Scientific American. Heidelberg: Springer Nature; 2019

[21] Pongsakornsathien N, Lim Y, Gardi A, Hilton S, Planke L, Sabatini R, et al. Senor networks for aerospace human-machine systems. Sensors. 2019;**19**:3-37. DOI: 10.3390/s19163465

[22] Rai B, Kaur J. Mental and physical workload, salivary stress biomarkers and taste perception: Mars Desert research station expedition. North American Journal of Medical Sciences. 2012;**4**:577-581

[23] Kaplan A, Haenlein M. On the interpretation, illustrations and implications of artificial intelligence. Business Horizons. 2019;**62**:15-25

[24] Dwivedi YK, Hughes L, Ismagilova E, et al. Artificial intelligence: Multidisciplinary perspectives on emerging challenges, opportunities, and agenda for research, practice and policy. International Journal of Information Management. 27 August 2019; Online. DOI: 10.1016/j.ijinfomgt.2019.08.002. Under press

[25] Culbert C, Boarnet M. AI applications for the space station. Robotics. 1988;**4**:35-40

[26] Sarris A. Antarctic station life: The first 15 years of mixed expeditions to the Antarctic. Acta Astronautica. 2017;**131**:50-54

[27] Agnoli S, Mancini G, Andrei F, Trombini E. The relationship between trait emotional intelligence, cognition and emotional awareness: An interpretative model. Frontiers of Psychology. 30 July 2019; Open access. DOI: 10.3389/fpsyg.2019.01711

[28] Fernandez-Berrocal P, Checa P. Editorial: Emotional intelligence and cognitive abilities. Frontiers of Psychology. 23 June 2016; Open access. DOI: 10.3389/fpsyg.2016.00955

[29] Eibl-Eibesfeldt I. L'homme programmé. Paris: Flammarion; 1976

[30] Tegmark M. President of the future of life science institute. Citation 2019. Available from: https://futureoflife.org/background/benefits-risks-of-artificial-intelligence/?cn-reloaded=1

[31] Tafforin C. Life at the Franco-Italian Concordia station in Antarctica for a voyage to Mars: Ethological study and anthropological perspectives. Antrocom. 2009;**5**:67-72

[32] Giner AF. Los Himbas: etnografia de una cultural ganadera de Angola y Namibia. Salamanca: Amarù Ediciones; 1992

[33] Kanas N, Saylor S, Harris M, Neylan T, Boyd J, Weiss DS, et al. High versus low crewmember autonomy in space simulation environments. Acta Astronautica. 2010;**67**:731-738

[34] Gushin V, Shved D, Vinokhodova A, Vasylieva G, Nitchiporuk I, Ehmann B, et al. Some psychophysiological and behavioral aspects of adaptation to simulated autonomous mission to Mars. Acta Astronautica. 2012;**70**:52-57

[35] Sharma G, Rai RN. Reliability modeling and analysis of environmental control and life support systems of space stations: A literature survey. Acta Astronautica. 2019;**155**:238-246

[36] Vinokhodova AG, Gushin VI. Study of values and interpersonal perception in cosmonauts on board of international space station. Acta Astronautica. 2014;**93**:359-365

[37] Tafforin C. In: Russomano T, Rehnberg L, editors. From the Individual to the Cultural Space Group. Into Space: A Journey of how Humans Adapt and Live in Microgravity. London: IntechOpen; 2018

[38] Anzieu D, Martin JY. La dynamique des groupes restraints. Paris: Presse Universitaire de France PUF; 2013

[39] Ming M, Custaud MA, Xu Z, Wang J, Yuan M, Tafforin C, et al. Multi-system adaptation to confinement during the 180-day Controlled Ecological Life Support System (CELSS) experiment. Frontiers in Physiology. 2019;**10**:1-22. DOI: 10.3389:fphys.2019.00575

[40] Sandal GM, Bye HH, Van de Vijver FJR. Personal value and crew compatibility: Results from a 105 days simulated space mission. Acta Astronautica. 2011;**69**:141-149

[41] Goel N, Bale TL, Neill Epperson C, Kornstein SG, Leon GR, Palinkas LA, et al. Effects of sex and gender on adaptation to space: Behavioral health. Journal of Women's Health. 2014;**23**:975-986

[42] Tafforin C, Giner Abati F. Cultural ethological as a new approach of interplanetary crew's behavior. In: Proceeding of the 65th International Astronautical Conference. Guadalajara, Mexico; September 2016

[43] Campan R, Scapini F. Ethologie: Approche systèmique du comportement. Bruxelles: De Boeck Université; 2002

Chapter 8

Psychosocial Aspects of a Flight to Mars

Radvan Bahbouh

Abstract

The first experiments modeling peoples' behavior during a long-term cosmic flight revealed the need for a more systematic monitoring of the development of the crew's mutual relationships, particularly in terms of collaboration and work-related communication. For this reason, in order to examine team dynamics, the sociomapping method was developed, which was first used in the HUBES 94 and ECOPSY 95 experiments. This method allows for an analysis and visualization of the continuous changes in communication and collaboration, including decreases in their quality and quantity. Sociomapping was used to monitor and analyze the communication and collaboration in simulations of flights to Mars in the Mars-105 and Mars-500 experiments. Based on the aforementioned experiments, it can be noted that statistically significant and nonrandom declines of the quantity and quality of team communication may occur during long-term missions, which may be related to changes in the team's performance. These changes are influenced by exterior stress factors, as well as cultural and linguistic differences and the length of the flight itself. In this chapter, the main findings of the experiment, as well as the resulting recommendations for a successful management of the psychological aspects of a flight to Mars, will be summarized.

Keywords: sociomapping, crew communication, psychological preparation, psychological support, debriefing, monitoring

1. Introduction

During the preparation of the first cosmic flights, the main question was their technical feasibility, which was first positively answered by the German rocket scientist Wernher von Braun. His technical feasibility study [1] had originally been an attachment to the science fiction story of a journey to Mars, which he wrote during his postwar internment [2]. In his story, he makes the crew members go through a 1-month isolation prior to their flight to Mars, in order to let this character screening test verify their ability to manage psychological distress [3]. When he modified this story for Collier's magazine edition, he writes about how states of tension and hatred appear among the crew members after several months, which could even have led to murder [2]. At this time, researchers were not only interested in physiological questions anymore but also in psychological aspects of extraatmospheric flights [4]. In the years 1956–1958, the first simulations of a stay in a cosmic cockpit took place, which included not only physiological but also psychological monitoring [5]. During one of these experiments, a decline in performance and mood was observed. Due to displays of hatred toward the researchers, the question arose whether to prematurely end the week-long isolation study [6]. The founding of the National Aeronautics and Space Administration (NASA) in 1958 allowed the extension of the experimental research program, including psychosocial aspects of long-term space flights. The psychological prerequisites for managing a cosmic flight were also considered important during the selection of the first astronauts in the Mercury project—besides an interview, as well as personality and performance psychological testing, candidates were sociometrically questioned, including whom of their colleagues they would commission to fly if they themselves could not and with whom they would most like to undertake the flight if the crew only had two members [5]. In his summarizing study, Christensen [7] reverted to the progress and results of this selection, to the existing interpersonal antagonisms, as well as to other findings. In explicit relation to the preparation for a flight to Mars, Christensen wrote about the need to research the impacts of cosmic flights on human behavior. Chambers [8] reached the conclusion in his summarizing study that psychological variables have a greater impact on the progress of simulated flights than physiological variables do. He also concluded that over time, motivation may decrease and that some crew members may become more irritable, even hostile. According to him, this can be prevented not only through a selection of suitable, i.e., stress-resistant, individuals, but also through training, which should include the team's isolation before the flight. In 1963, the Institute of Biomedical Problems (IMBP) was founded in the Soviet Union, whose purpose was to deal, among other things, with the research of the conditions of a flight to Mars [9]. In 1964, the first experiment took place in the IMBP—a 120-day-long stay of a three-member crew in a module imitating a spaceship designed for a flight to Mars [10, 11]. Following up, an experiment of a 1-year isolation of a three-member crew took place in 1967–1968 [11]. As one of the participants of this 1-year experiment later indicated [12], it turned out that "the problem of the crew's psychological tolerance is one of the key problems of the medicinally psychological assurance of lengthy cosmic flights." Fraser [13] analyzed 60 confinement studies and also reached the conclusion that feelings of anger and animosity, directed either at other crew members or at the researchers and other outside persons they communicate with, occur frequently. Haythorn [14] indicated in his summarizing study that interpersonal relationships do not always alleviate stress but can also produce it. Furthermore, he notes that little is known about the behavior of isolated groups in stress situations, due to the lack of longitudinal studies. As Suedfeld [15] later pointed out, it is possible to deduce, to a certain extent, the behavior of people on long-term cosmic missions from early terrestrial and marine exploration voyages, which also illustrate the importance and vulnerability of mutual interpersonal relationships. Kanas and Fedderson [16] conducted their review bearing in mind the fact that the mission to Mars will be a long-term one. Among other things, they point out that in the longer term, restrictions of interpersonal contact may appear. Like other similarly oriented research, they recommend reducing the risks with a suitable selection of people and with activities compensating the monotony and sensory deprivation. Vinograd's work [17] was a synoptical summarization in its time, comprised of 14 studies simulating cosmic flights but also of comparable studies of submarine crew members, teams in arctic or military environments, or in other situations of isolation. An analysis of individual studies reveals not only the importance of mutual relationships but also the lack of tools for a continuous (quantitative) capturing of their changes. In 1975, the newly founded European Space Agency (ESA) joined the other research institutions in their coverage of cosmic research (including psychosocial questions).

2. Deepening of the research on psychosocial factors

During the 1970s, it was proven during real cosmic flights that relationship and communication issues can have grave consequences. In 1974, the first cosmic strike of a crew led by William Pogue took place, who reasons in his autobiography [18] that the strike happened due to the demanding work program, which did not include time for rest. The exhaustion culminated in a conflict between the control center and the astronauts, to which the astronauts reacted with a 24-hour silence. In 1975, the astronauts themselves pointed out the importance of the psychological aspect of space flights [19]. In 1976, a premature abortion of the mission of the two-member crew of Volynov and Zholobov occurred, which was later attributed to the demanding work conditions, the worsening psychological state of the crew, but also to the serious conflicts between the two astronauts [20, 21]. Almost 10 years later, Harrison and Connors [22] suggested lowering the psychological and interpersonal vulnerability of the team not only by a suitable selection of crew members but also through training in group dynamics and by offering psychological support. They request that the anecdotal testimonies of the importance of these factors be examined by systematic scientific research. In cooperation with the IMBP, the Štola 88 experiment was conducted in the former Czechoslovakia, in which the comparison of two teams simulating a flight to Mars showed how differently the communication can develop during isolation depending on the composition of the team and leadership type [23–25]. In this 23-day experiment (structured for the crew as thirty 18-hour days), it was shown, among other things, that part of the tension manifested itself in a deterioration of the communication with the control center [24, 25]. In a later debriefing, the presence of a woman in one of the teams was mentioned as positive [26]. Related to the renewed considerations of a journey of six (or more) astronauts to Mars, Kanas [27] points to the possible impact of psychological, psychiatric, and interpersonal factors on the safety and success of such a mission. He mentioned interpersonal tension, a continuously decreasing team cohesiveness, the need for privacy, and the leadership's contradicting focus on the task and on emotions. In the 1990s, under the tutelage of the European Space Agency, a 30-day experiment with a six-member crew, Isolation Study for European Manned Space Infrastructure (ISEMSI-90), was carried out [28], as well as a 60-day experiment with a four-member crew, Experimental Campaign for the European Manned Space Infrastructure (EXEMSI-92) [29]. To capture the relationship dynamics, the Systematic Multiple Level Observation of Groups (SYMLOG) method [30, 31] was utilized, for example, as well as an analysis of spatial behavior [32, 33] and an analysis of the communication with the control center [34, 35]. During a summary of the EXEMSI experiment's results [36], it was noted that while no conflicts occurred within the team, this came at the expense of suppressing affection and a more rigid functioning of the team. Cazes and his colleagues believe that it was possible to maintain such a communication thanks to the experiment's relatively short duration and to the absence of any real risk. They consider this type of behavior inadequate (even dangerous) for a real space flight. They also point out that the team's cohesiveness was maintained by using the management as a scapegoat, as demonstrated by the criticism of the ground crew during crisis situations. For this reason, Cazes et al. [36] have doubts whether the harmony presented in sociometric tests was real or apparent. An important stimulus for the research of communication in the 1990s was the analyses proving that a majority of accidents in aviation are caused by human factors [37]. As it later turned out, after the introduction of standardized communication rules (the crew management system), accidents caused by human factors decreased significantly [38]. In 1994, a joint project between the ESA and IMBP was carried out, the Human Behavior in Extended Spaceflight (HUBES) experiment, in which a 30-member crew was isolated for 135 days. In this experiment, physiological variables were observed in addition to the crew's communication. Sociometric tests were included, as well. The sociomapping method, which had been tested for a year on military units of the Czech army [39], supplemented the data collection and will be presented in more depth in the following chapter. In 1996, a 30-day stay of a four-member crew took place as part of the Lunar-Mars Life Support Test Project [40]. Even though the team dynamic was evaluated as ideal during the debriefing of this experiment, one of the recommendations was for the teams to be briefed more and sensitized to the psychological aspects of the experiments. In the summary of the subsequent 91-day experiment that took place in the Lunar-Mars Life Support Test Project in 1997, its leader mentions that in the early phases of the project, miscommunications occurred between the control center and the crew [41], to which it was necessary to react with an increased emphasis of the "overall team-integration approach," which also included members of the management. Holland a Curtis [42] pointed out in their summary of the results of NASA studies within the Lunar-Mars Life Support Test Project that extending the length of the mission increases the significance of psychological factors and thus psychological activities, as well, which are meant to ensure the success of the mission [43]. Among those, they mention training, briefing, in-mission tracking, and prospective interventions. They also allude to the importance of communication with the family and other close people outside of the crew. As Galarza and Holland [44] propose, the fact that teamwork and the ability to get along with the team are critical competencies for long-term flights should be reflected in the development of tools and procedures for the selection of people but also in their trainings and in-flight support. Despite the knowledge that tensions may rise during periods of isolation and partial communication isolations of certain team members may occur, the next experiment conducted in the IMBP, Simulation of Flight of International Crew on Space Station (SFINCSS), was carried out without a continuous monitoring of the communication. Furthermore, it utilized subjective evaluation scales and did not provide an appropriate training on group dynamics, which would include cultural and gender aspects, too (a woman took part in the experiment, unlike the previous HUBES and EKOPSY experiments). The experiment had to be terminated prematurely due to an argument between two astronauts, resulting in a physical altercation, an allegation of harassment, and due to the explicit request of a Japanese crew member to be able to leave the shuttle [45, 46]. Despite its failure, this experiment was useful, as it pointed out what consequences an underestimation of a continuous monitoring of communication and its subsequent interventions can have. It also demonstrated that it is necessary to pay attention to linguistic and cultural aspects. Morphew [47] refers to personal conversations with astronauts Jdanov and Atkov to point out that the astronauts themselves consider psychological and psychosocial aspects among the most critical problems of long-term flights, based on their own experience. Among the most significant psychosocial stress factors, Morphew mentions the high demands on team coordination, tensions between the crew members and control center, the forced contact with other crew members, the lack of contact with the family, cultural differences, and other factors, such as differences in gender, personalities, and others. Other non-psychosocial factors, such as high or low workloads, lack of privacy, space adaptation sickness, and of course the permanent life-threatening dangerous environment, must be considered, as well. Morphew [47] concludes that the US cosmic program considers psychological factors critical for increasing the safety and ensuring the success of the mission. In 2004, Manzey states that the research on human behavior during long-term missions is still insufficient to estimate and reduce specific risks associated with a long-term journey to Mars. In this context, he mentions that it is necessary to pay attention not only to an individual adaptation and performance but also to the interactions among the crew members and methods of psychological measures [48, 49].

3. Sociomapping the communication of the crew on a flight to Mars

The sociomapping method allows to visually express mutual proximities (and distances) between individual teammates, military units, and crews [39, 50–52]. From a mathematical perspective, proximity is the degree of membership into the fuzzy set of people close to a specific member of the team. Various operational definitions are being used for the degree of membership, depending on the situation. Mutual proximity can be defined by the time spent on joint conversations, the volume of text or information, the average physical proximity, and many other characteristics. Most commonly, scales evaluating the mutual communication or cooperation in a given timeframe from a quantity and quality standpoint are used [52]. Such operationalized values of mutual proximity do not have to be symmetrical for two team members. Data about the validity, reliability, and time dependability is known and constantly being supplemented [52–55]. During the creation of sociomaps, the order of the closest to the furthest colleague is correlated for each team member in terms of the spatial distance ranked by the closest to the furthest according to the degree of membership. The final sociomap is created by maximizing the average Spearman correlation coefficients calculated for each team member [51, 52, 56]. During sociomapping, the average values of the scales are being monitored using the control chart method, which allows to capture significant deviations over time [52, 57, 58]. In the HUBES experiment, it turned out [23] that sociometric tests were not sensitive enough, whereas the scales evaluating the cooperation allowed to capture the gradual development, which consisted in one crew member separating from the other two with a simultaneous decrease in communication (substantiated by analyses of actual communication). The aggregated score expressing the degree of subjectively and physiologically captured stress grew over time, particularly in the final quarter of the experiment [23, 50]. As (not only) the HUBES experiment showed, traditional sociometrical procedures consisting in the selection of the remaining crew members are not very suitable due to their lack of sensitivity to continuous changes, which is particularly important in long-term missions. For this reason, sociomapping was also used in the 90-day experiment ECOPSY-95, in which a three-member crew was expanded by another three-member crew over the course of the experiment. Thanks to sociomapping, it was possible to capture how both crews interconnected from a communication standpoint, particularly thanks to the communication between the two crew leaders [51, 59]. After the departure of the second crew, the original three-member crew returned to the initial composition, while one of the members remained relatively separated from the communication perspective. In the Mars-105 experiment, a flight to Mars was simulated throughout a 105-day stay of a six-member crew in a module of the MIR ship [52]. To monitor the communication, subjective scales were used again, including a five-point evaluation of changes in the communication frequency with individual crew members, for example, the communication in the last previous weeks decreased significantly – decreased slightly – stayed the same – increased slightly – increased significantly. In addition, it included a five-point evaluation of the required optimal change in communication frequency (a wish for a significant decrease – slight decrease – maintenance of the current state – slight increase – significant increase), a percentage evaluation of the quality of cooperation with individual team members (0–100% scale), as well as an overall assessment of the team's performance and atmosphere (0–100% scale). The scales were supplemented by a question about the frequency of misunderstandings. All questions concerned the period from the previous measurement (i.e., approximately the prior 2 weeks). The following graph indicates the average values of the five-point scales of the perceived communication changes converted to the interval <0;1> for seven specific dates indicated by the administration (the whole experiment took place between March 31, 2009, and July 14, 2009). The value of 0.5 corresponds to constant, or stabilized, in terms of communication frequency (**Figure 1**).

Even though the communication frequency stabilized throughout the experiment, it is possible to capture significant changes in the percentage scales of the assessment of the mutual communication's quality, which match the fluctuations of the average percentage scales of the overall evaluation of the team's performance and atmosphere (**Figure 2**).

In the mutual assessment of the communication quality and in the overall evaluation of the team's performance and atmosphere, the minimum was reached in the administration on June 13, 2009, i.e., about two thirds into the experiment. The assessment of the performance and quality of communication were even lower at this point than it had been during the first administration. Furthermore, for this period, the highest number of misunderstandings was reported (only two out of six people indicated that no misunderstandings occurred in this period). In a personal account, the crew's leader commented on the worsening atmosphere, performance, and quality of communication. Among other factors, he considered fatigue, as well as the suggested improvement measures, which the crew submitted to the control center and which were not accepted according to their expectations, to have played a role. The deterioration of communication quality, captured on multiple levels, proves that it is possible to detect significant changes using a scaled assessment, even though the scales are test–retest reliable. The median of the test–retest correlation in this experiment was 0.8 [52]. While the average scale values may point to the fact that the communication overall is deviating from its norm, the sociomaps offer insight on an individual team member level. The following figures show sociomaps of the crew (including the displayed position of the control center) dated May 2, 2009, and May 9, 2009 (in this case, the time passed between the administration was only 7 days) (**Figures 3** and **4**).

As the description of sociomapping above implies, the closer two crew members are to each other, the more they communicate together. The whole team can be

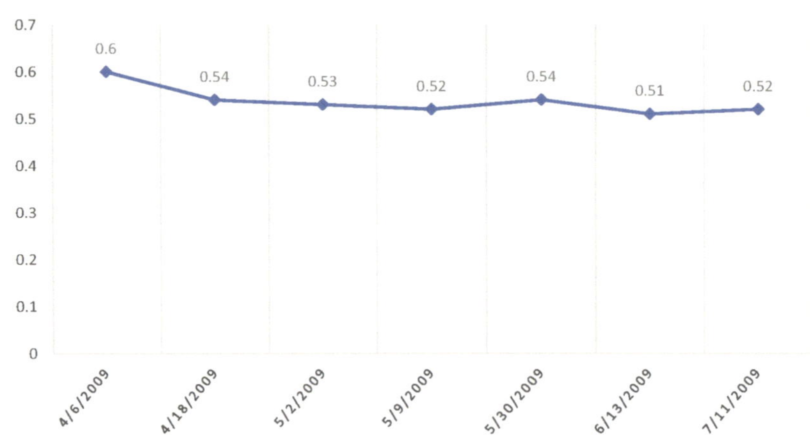

Figure 1.
Average change in frequency of communications.

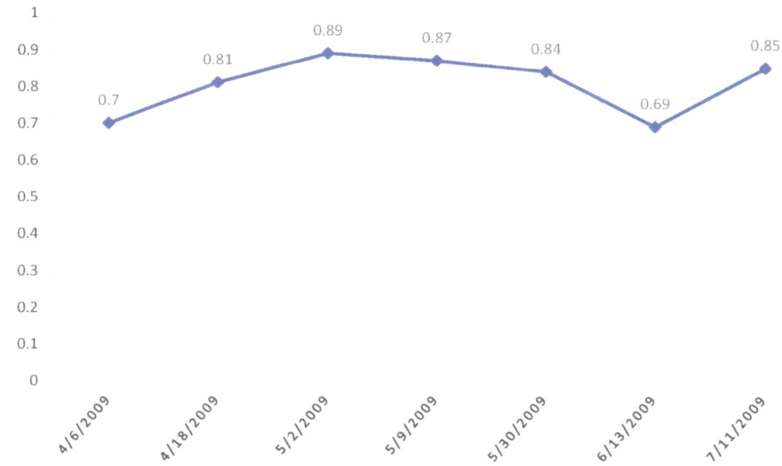

Figure 2.
Average quality of cooperation.

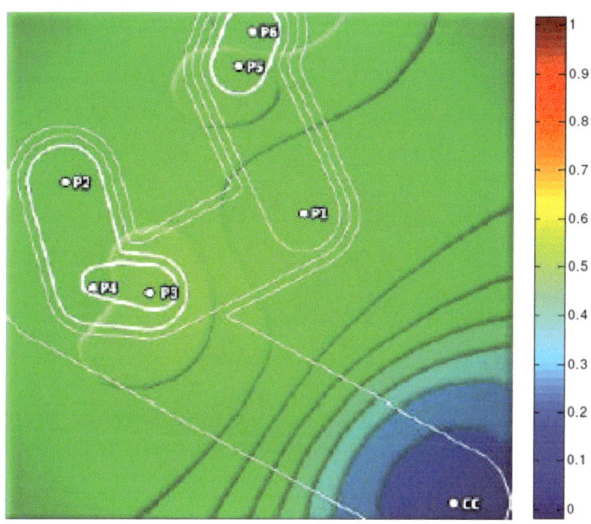

Figure 3.
Sociomap of Mars-105 crew (May 2, 2009).

divided into subsets according to their mutual adhesion [50–52]. The blue arrows indicate a desired increase in communication (compared to the current level), while red arrows signify a wish for the decrease of communication. Colors stand for the average frequency of current communication for each crew member and control center (CC).

The situation changed over the course of 7 days—person P2 is more connected with persons P3 and P4. The basis of the second subgroup is primarily the close relationship between P5 and P6, the only foreign (non-Russian) crew members. Following the Mars-105 experiment, the longest experiment to date took place, lasting 520 days—Mars-500, during which communication was also continuously monitored. Mars-500 simulated a flight to Mars in the full scale of its estimated 520-day duration [57, 60]. Sociomaps were continuously sent to the control center. The option of notifying the center about unusual or unexpected situations for possible intervention purposes was not utilized in this case, as the communication was relatively stable throughout the experiment, both in terms of quantity and quality.

The entire communication progress may be viewed dynamically based on individual sociomaps created in regular 14-day intervals (36 administrations in total). The simulated landing on Mars occurred in February 2011, when the team was separated into two subteams—a landing and orbital team. Control center is included (Video 1 can be viewed at https://bit.ly/2ENYzja).

Even though the average correlation between the values of subsequently occurring sections was high (the average scale value of the test–retest 14-day correlation was 0.785 for the current communication and 0.843 for the optimal communication), a decrease could be observed in some cases. The test–retest correlations for the 16th and 17th collection, which took place before and after the landing, were −0.005 and 0.117, respectively [60]. Since each of the six crew members expressed their opinion about the remaining five members in the relational questions, 30 values were gathered in total in the individual collections. The following graph (**Figure 5**) shows the differences in absolute values between the actual degree of communication expressed on a five-point scale and the optimal communication value.

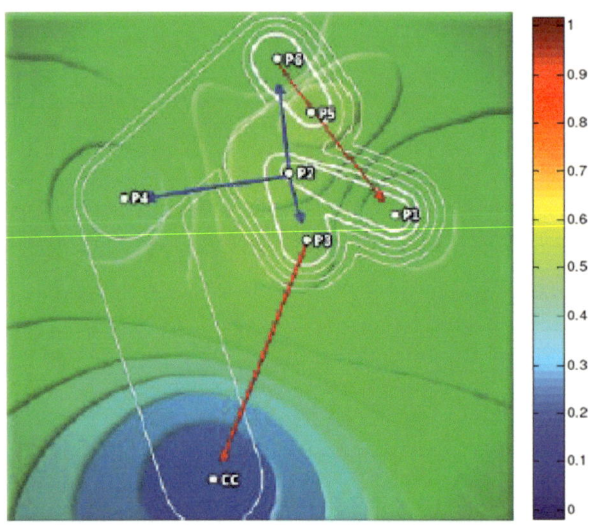

Figure 4.
Sociomap of Mars-105 crew (May 9, 2009).

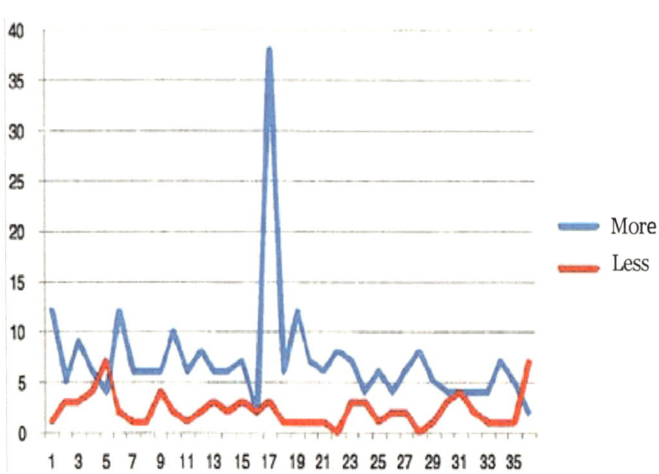

Figure 5.
Differences between current and optimal communication in Mars-500 experiment.

Figure 6.
Aggregated sociomap of Mars-500 experiment.

The most visible change occurred during the landing period. The occurrence of misunderstandings negatively correlated with the overall level of current communication ($r = -0.33$), while it positively correlated with the expressed wish for a communication change ($r = 0.42$). Besides creating partial sociomaps, it is possible to aggregate the sociomap data, as well as that from the derived sociomap, for an extended period. That way, it is even possible to aggregate all 36 data collections into one sociomap expressing the interconnectedness of the individual crew members (**Figure 6**).

The main linking factor among the crew members is person U, who is closer with pair R and S. The triplet U, R, and S, as well as G, E, and N, forms the Russian and non-Russian team subgroup, respectively. Communication within these subgroups is statistically higher than communication between (people from) both subgroups ($p = 0.002$, $d = 1.28$), which accentuates the importance of understanding linguistic and cultural aspects of communication and cooperation in crews. The Mars-500 experiment showed that this way of monitoring is sensitive enough (despite the high test–retest reliability of the scales) and that the crew can handle communication even in such long isolation periods without long-term deterioration. Based on our experience and research, we assume that monitoring alone can sensitize the crew to the importance of communication and instigate possible attempts to change this communication.

The advantages of monitoring communication using sociomapping continue to be examined in other experiments simulating space flights [61–63]. The interventions' success is being researched experimentally and quasi-experimentally within work teams [53, 64–68]. The main source of our findings is all work teams we supported with sociomapping, some of which were being examined over the course of more than 3 years [52]. The second significant source is the usage of sociomaps in the Czech Army, where this method has been used in combat teams for 25 years [69].

4. Conclusion

From the experiments to date, which directly or indirectly simulate a flight to Mars, it can be deduced that psychosocial factors are critically important and their underestimation may even lead to a failure of the whole mission. Besides selecting suitable, stress-resistant individuals during the preflight phase, it is recommendable for the crew members to meet well in advance and for them to spend some time together in isolation before the actual flight. During this phase, it would be appropriate for the crew to be briefed with the issue at hand and with the usual development of the group dynamic, the methods that will help them capture it, and with the subsequent procedures of formulating a contract about what can be improved based on this data. We also recommend paying attention to a sensitization on cultural, linguistic, and gender aspects during the training and briefing prior to the actual flight. We also suggest for the team to familiarize itself with the way of conducting the debriefing, which improves its communication and performance. This is based not only on the meta-analysis, which showed a significant increase of the performance of teams that conduct debriefings [70], but also from the derived recommendations aimed specifically at space flights [71]. In the view of the communication delay between Earth and Mars, it is necessary to support the teams by leading them to conduct the debriefing based on an automated, structured protocol. For example, this could be sociomapping followed by a debate about how the communication is going and how it might be continuously improved. Throughout the mission, we recommend a regular monitoring of the communication with a possible discussion on improvement options. From time to time, the team "on Earth" should be included in such a discussion, so that this intervention can help strengthen the mutual ties and reduce the risks of possible displacement. Besides a debriefing, providing support or an intervention may also be considered in more difficult times. We also unequivocally suggest that the contact with close friends and relatives be available, which will decrease the sense of isolation. Monitoring communication using the sociomapping method will continue to be a part of the experiment in the Sirius project. After the Sirius 2017 [72] and Sirius 2019 experiments, the utilization of sociomapping is planned in the Sirius 2020 and 2021 experiments, as well. The team led by Kateřina Bernardová is now preparing not only a methodology of the measurement itself but also a methodology of team dynamic training and subsequent interventions similarly to the way the Czech Army has been using the system for the last 20 years [69].

Acknowledgements

I would like to thank the pioneer of Czech space research, military pilot, and sociologist Col. PhDr. Jaroslav Sýkora, as well as psychologist Col. PhDr. Kateřina Bernardová, founder of the Department of Psychosocial Services in the Czech Army, with whom I had the honor to collaborate for over 25 years. Col. PhDr Kateřina Bernardová was using sociomapping for interventions in combat units for over 20 years and now is leading Sirius project. I would like to thank PhDr. Eva Rozehnalová, PhD, for the collaboration on psychometric research of sociomapping, and RNDr. Cyril Höschl, PhD, for translating the idea of sociomapping into functioning algorithms and programs.

Author details

Radvan Bahbouh
Department of Psychology, Faculty of Arts, Charles University, Prague, Czech Republic

*Address all correspondence to: radvan.bahbouh@ff.cuni.cz

References

[1] von Braun W. The Mars Project. Urbana, IL: University of Illinois Press; 1953. Available from: https://archive.org/stream/TheMarsProject-WernherVonBraun1953

[2] Portree DS. Humans to Mars: Fifty Years of Mission Planning, 1950-2000. Washington, DC: NASA History Division; 2001. Available from: https://history.nasa.gov/monograph21.pdf

[3] von Braun W. The Mars Project. A Technical Tale [Internet]. Burlington: Apogee Books; 2006. Available from: https://archive.org/details/ProjectMars

[4] Armstrong HG, Haber H, Strughold H. Aero medical problems of space travel. The Journal of Aviation Medicine. 1949;**XIX**:383-417

[5] Swenson L, Grimwood J, Alexander C. This new ocean: A history of project Mercury. NASA Special Publication-4201 in the NASA History Series [Internet]; 1989. Available from: https://www.hq.nasa.gov/office/pao/History/SP-4201/toc.htm

[6] Hauty GT. Human performance in space. In: Gantz KF, editor. Man in Space: The United States Air Force Program for Developing the Spacecraft Crew. New York: Duell, Sloan and Pearce; 1959. pp. 84-108

[7] Christensen JM. Psychological aspects of extended manned space flight. In: Grether WF, editor. Aerospace Medical Division, Air Force System Command, Wright-Patterson Air Force Base. Ohio: Air Force Aerospace Medical Research Lab Wright-Patterson AFB; 1963

[8] Chambers RM. Isolation and disorientation. In: Hardy JD, editor. Physiological Problems in Space Exploration. Springfield Illinois: Thomas Publisher; 1964. pp. 231-292

[9] Gazenko OG. Milestones of space medicine development in Russia (establishment and evolution of the Institute of Biomedical Problems). Journal of Gravitational Physiology. 1997;**4**(3):1-4

[10] Bugrov V. Marsianskij projekt S.P. Koroljova – perspektivnaja cel Rossijskoj kosmonavtiki. Aktualnyje problemy aviacionnych i aerokosmičeskich sistem: processy, modeli, eksperiment. 2015;**1**(20):102-124

[11] Toufar P. NEK – laboratoř pro Mars i pro orbitální stanice. In: Šolcová I, Stuchlíková I, Guščin VI, editors. MARS-500 Fakta a postřehy ze simulovaného letu na rudou planetu. Praha: Academia; 2014. pp. 61-78

[12] Manovcev GA. Od ročního pobytu v "pozemním hvězdoletu" k Marsu-500. In: Šolcová I, Stuchlíková I, Guščin VI, editors. MARS-500 Fakta a postřehy ze simulovaného letu na rudou planetu. Praha: Academia; 2014. pp. 79-82

[13] Fraser TM. Confinement and free-volume requirements. Space Life Sciences. 1968;**1**(2-3):428-466

[14] Haythorn WW. Interpersonal stress in isolated groups. In: McGrath JE, editor. Social and Psychological Factors in Stress. New York: Rinehart and Winston; 1970

[15] Suedfeld P. Historical space psychology: Early terrestrial explorations as Mars analogues. Planetary and Space Science. 2010;**68**(4):639-645

[16] Kanas NA, Fedderson WE. Behavioral, Psychiatric, and Sociological Problems of Long-Duration Space Missions. Houston, Texas: Manned Spacecraft Center; 1971. [NASA TM X-580]

[17] Vinograd SP. Studies of social group dynamics under isolated conditions. Objective summary of the literature as it relates to potential problems of long duration space flight [Internet]. 1974 December [cited 13 December 2019]. Available from: https://ntrs.nasa.gov/search.jsp?R=19750007236

[18] Pogue W. But for the Grace of God: An Autobiography of an Aviator and Astronaut: Golden. Golden, CO: ReAnimus Press/CreateSpace Independent Publishing Platform; 2016

[19] Leonov AA, Lebedev VI. Psychological Problems of Interplanetary Flight (NASA-TT-F-16536). Washington DC: NASA; 1975

[20] Burgess C, Hall R. The First Soviet Cosmonaut Team: Their Lives, Legacy, and Historical Impact. Berlin: Springer; 2009

[21] Šamárek O. Kritické momenty kosmonautiky, 18.díl [Internet]. 2013. Available from: https://www.kosmonautix.cz/2013/01/kriticke-momenty-kosmonautiky-18-dil/

[22] Harrison AA, Connors MM. Psychological and interpersonal adaptation to Mars missions. In: The Case of Mars II. San Diego: Univelt; 1985. pp. 643-654

[23] Sýkora J, Šolcová I, Bahbouh R, et al. Group dynamics and tension analysis in simulated extended spaceflight settings. In: Advances in Space Biology and Medicine: HUBES – 135 Days of MIR Simulation Study. Paris: ESA; 1996

[24] Šolcová I. Poznatky o psychické odolnosti malých skupin izolovaných v náročných podmínkách. Československá Psychologie. 2008;**52**(3):253-264

[25] Šolcová I. Poznatky o odolnosti malých skupin izolovaných v náročných podmínkách. In: Šolcová I, Stuchlíková I, Guščin VI, editors. Mars-500 Fakta a postřehy ze simulovaného letu na rudou planetu. Praha: Academia; 2014

[26] Sýkora J, Šolcová I, Dvořák J, Polánková M, Tomeček A. Women's place in space exploration. Space Policy. 1996;**12**(3):161-165

[27] Kanas N. Psychological, psychiatric, and interpersonal aspects of long-duration space missions. Journal of Spacecraft and Rockets. 1990;**27**(5):457-463

[28] Bonting SL et al. European isolation and confinement study. Advances in Space Biology and Medicine. 1993;**3**:1-273

[29] Bonting SL et al. Exemsi: The second european simulation of a long-duration manned space mission. Advances in Space Biology and Medicine. 1996;**5**:1-401

[30] Rosnet E, Bachelard C, Cazes G, Le Scanff C, Rivolier J. A comparative study between different psychological approaches during an ESA space simulation. SAE Technical Paper; 1994

[31] Sandal GM, Værnes R, Ursin H. Crew compatibility and interaction. Advances in Space Biology and Medicine. 1996;**5**:213-231

[32] Tafforin C. Chapter 7 ethological analysis of spatial behavior. In: Bonting SL, editor. Advances in Space Biology and Medicine [Internet]. Goor, Netherlands: Elsevier; 1993 [cited 13 December 2019]. pp. 81-94 (European Isolation and Confinement Study; Vol. 3). Available from: http://www.sciencedirect.com/science/article/pii/S1569257408600982

[33] Tafforin C. Chapter 15 ethological analysis of crew member behavior: Distances, orientations, and postures.

In: Bonting SL, editor. Advances in Space Biology and Medicine [Internet]. Vol. 5. Goor, Netherlands: Elsevier; 1996. pp. 263-281. [cited 13 December 2019]. Available from: http://www.sciencedirect.com/science/article/pii/S1569257408600647

[34] Gushin VI, Kolinitchenko TB, Efimov VA, Davies C. Psychological evaluation and support during EXEMSI. Advances in Space Biology and Medicine. 1996;**5**:283-295

[35] Gushin VI, Yusupova AK, Shved DM, Shueva LV, Vinokhodova AG, Bubeev YA. The evolution of methodological approaches to the psychological analysis of the crew communications with Mission control center. REACH. 2016;**1**:74-83

[36] Cazes C, Rosnet E, Bachelard C, Le Scanff C, Rivolier J. Group dynamics during the EXEMSI isolation study. Experimental campaign for the European manned space infrastructure. Advances in Space Biology and Medicine. 1996;**5**:245-262

[37] Weigmann DA, Shappell SA. Human factors analysis of postaccident data: Applying theoretical taxonomies of human error. The International Journal of Aviation Psychology [Internet]. 1997;**7**(1):67-81. DOI: 10.1207/s15327108ijap0701_4

[38] Lehrer J. How We Decide. New York: Houghton Mifflin Harcourt; 2010

[39] Bahbouh R. Sociomapování (Sociomapping). Prague: Gema Art; 1994

[40] Ming DW, Hurlbert KM, Kirby G, Lewis JF, ORear P. Lunar-Mars Life Support Test Project. Phase 2; Human Factors and Crew Interactions; 1997 retrieved 2019. Available from: https://ntrs.nasa.gov/archive/nasa/casi.ntrs.nasa.gov/19970025650.pdf

[41] Packham NJ. The Lunar-Mars life support test project- the crew perspective. In: Isolation: NASA Experiments in Closed-Environment Living- Advanced Human Life Support Enclosed System Final Report [Internet]. Vol. 104. 2002 retrieved 2019. pp. 17-34. Available from: https://lsda.jsc.nasa.gov/books/ground/1.3Crewmembers.pdf

[42] Holland AW, Curtis K. Operational psychology countermeasures during the lunar-Mars life support test project. Life Support & Biosphere Science. 1998;**5**(4):445-452

[43] Holland AW. Opcrational psychology countermeasures during the Lunar-Mars life support test project. (LMLSTP3.5) [Internet]. 2019 [cited 13 December 2019]. Available from: https://lsda.jsc.nasa.gov/Experiment/exper/927

[44] Galarza L, Holland AW. Critical astronaut proficiencies required for long-duration space flight. SAE Technical Paper; 1999

[45] Inoue N, Matsuzaki I, Ohshima H. Group interactions in SFINCSS-99: Lessons for improving behavioral support programs. Aviation, Space, and Environmental Medicine. 2004;**75**(7):C28-C35

[46] Sandal GM. Culture and tension during an international Space Station simulation: Results from SFINCSS'99. Aviation, Space, and Environmental Medicine. 2004;**75**(7):C44-C51

[47] Morphew ME. Psychological and human factors in long duration spaceflight. McGill Journal of Medicine [Internet]. 2001;**6**(1):74-80. Available from: http://citeseerx.ist.psu.edu/viewdoc/download?doi=10.1.1.484.9692&rep=rep1&type=pdf

[48] Kanas MD. Space Psychology and Psychiatry. Dordrecht: Kluwer Academic Press; 2003

[49] Manzey D. Human missions to Mars: New psychological challenges and research issues. Acta Astronautica. 2004;**55**(3-9):781-790

[50] Bahbouh R. Využití sociomapování u malých sociálních skupin. (Use of sociomapping in small social groups) [Master's thesis]. Prague: Department of Psychology, Charles University in Prague; 1996

[51] Bahbouh R. Sociomapování (Sociomapping) [PhD thesis]. Prague: Department of Psychology, Charles University in Prague; 2004

[52] Bahbouh R. Sociomapping of Teams. Prague: QED GROUP; 2012

[53] Bahbouh R, Rozehnalová E, Děchtěrenko F. Dependability of group communication sociomapping. In: Lasker G, Bahbouh R, editors. Sociodiagnostics and Sociomapping. Vol. II. Tecumseh, Ontario: The International Institute for Advanced Studies in Systems Research and Cybernetics; 2013. pp. 4-10

[54] Rozehnalová E, Bahbouh R. Effectiveness of sociomapping intervention I: Theoretical background. In: Lasker G, Bahbouh R, editors. Sociodiagnostics and Sociomapping. Vol. II. Tecumseh, Ontario: The International Institute for Advanced Studies in Systems Research and Cybernetics; 2013. pp. 11-16

[55] Kubík R, Bahbouh R. The relationship between satisfaction with team communication and evaluation of communication with individual team members. In: Bahbouh R, Lasker G, editors. Sociodiagnostics and Sociomapping. Vol. 3. Tecumseh, Ontario: The International Institute for Advanced Studies in Systems Research and Cybernetics; 2014. pp. 46-50

[56] Höschl C. Vizualizace sociomap [Master's thesis]. Prague: Charles University; 2006

[57] Bahbouh R, Děchtěrenko F. Sociomapping of communication changes during simulations of flights to Mars. In: Bahbouh R, Lasker G, editors. Sociodiagnostics and Sociomapping. Vol. III. Tecumseh, Ontario: The International Institute for Advanced Studies in Systems Research and Cybernetics; 2014. pp. 34-40

[58] Shewart W. Statistical Method from the Viewpoint of Quality Control: by Walter A. Shewhart. New York: Dover; 1986.

[59] Sýkora J, Nechaev AP, Šolcová I, Bahbouh R, Konopásek P. Izučenije gruppovoj dinamiki v uslovijach 90-sutočnoj izoljaci. (evaluation of the dynamics of in-group strain during long isolation). Aviakosmičeskaja i Ekologičeskaja Medicina. 1997;**34**:11-15

[60] Bahbouh R, Sněhotová J, Děchtěrenko F, Sýkora J. Sociomapování komunikace posádky v experimentu Mars 500. In: Guščin V, Stuchlíková I, Šolcová I, editors. Mars-500 Fakta a postřehy ze simulovaného letu na rudou planetu. Praha: Academia; 2014. pp. 147-163

[61] Davidová L. Sociomapping of a student crew during a simulation of a flight to Mars. In: Bahbouh R, Lasker G, editors. Sociodiagnostics and Sociomapping. Vol. IV. Tecumseh, Ontario: The International Institute for Advanced Studies in Systems Research and Cybernetics; 2015. pp. 15-26

[62] Kubík R, Vyšatová K. Trust, communication, and quality of cooperation within a crew during the "Expedice vesmír 2014" experiment. In: Bahbouh R, Lasker G, editors. Sociodiagnostics and Sociomapping. Vol. IV. Tecumseh: The International Institute for Advanced Studies in Systems Research and Cybernetics; 2015. pp. 27-37

[63] Vraný M, Bezchlebová LL. The relationship between the volatility

of respondents¨ratings and team performance. In: Bahbouh R, Lasker G, editors. Sociodiagnostics and Sociomapping. Vol. IV. Tecumseh: The International Institute for Advanced Studies in Systems Research and Cybernetics; 2015. pp. 38-44

[64] Fabianová I. Efektivita týmového koučování [PhD thesis]. Prague: Charles University; 2019

[65] Kovaříková J, Rozehnalová E, Bahbouh R. Effectiveness of sociomapping intervention II: Case studies. In: Lasker G, Bahbouh R, editors. Sociodiagnostics and Sociomapping. Vol. II. Tecumseh, Ontario: The International Institute for Advanced Studies in Systems Research and Cybernetics; 2013. pp. 17-22

[66] Rozehnalová E. Reliabilita a validita sociomapování komunikace [PhD thesis]. Prague: Charles University; 2013

[67] Tetour V. Efektivita intervence sociomapování u vybraných charakteristik pracovních týmů [Master's thesis]. Prague: Charles University; 2019

[68] Vraný M, Bahbouh R. Sociomapping of the mutual familiarity of team members and its increase. In: Bahbouh R, Lasker G, editors. Sociodiagnostics and Sociomapping. Vol. IV. Tecumseh: The international institute for advanced studies in systems research and cybernetics; 2015. pp. 45-51

[69] Bernardová K. Using sociomapping with the Czech Army. In: Bahbouh R, Rozehnalová E, Sailerová V, editors. New Perspectives of Psychodiagnostics. Prague: QED GROUP; 2012. pp. 71-82

[70] Tannenbaum SI, Cerasoli CP. Do team and individual debriefs enhance performance? A meta-analysis. Human Factors [Internet]. 2013;**55**(1):231-245. DOI: 10.1177/0018720812448394

[71] Salas E, Tannenbaum SI, Kozlowski SW, Miller CA, Mathieu JE, Vessey WB. Teams in space exploration: A new frontier for the science of team effectiveness. Current Directions in Psychological Science. 2015;**24**(3):200-207

[72] Tefelnerová P. Vývoj struktury a dynamiky vztahů a vazeb mezi členy posádky v průběhu simulace kosmického letu [Master's thesis]. Prague: Charles University; 2019

Chapter 9

Astrosociology on Mars

Jim Pass

Abstract

This chapter focuses on the importance and need to focus much more strongly on the social and behavioral sciences, humanities, and arts regarding the study of space issues, and Mars issues specifically. The particular focus here deals with the vital need to balance planning for Martian settlements between the two major branches of science [i.e., (1) the physical and natural sciences plus the STEM disciplines and fields and (2) the social sciences, humanities, and arts]. The current focus on settling Mars is on transportation issues and physical infrastructure construction of habitats or cities. In contrast, the arguments made here support the idea that planning for Martian settlements must include the "other" branch of science, and this extends to the monitoring of social interactions and other forces associated with the complexities generated by a growing population existing in an isolated and dangerous ecosystem on another planet. On Earth, the social sciences are vital and they will be on Mars as well. Astrosociology is a multidisciplinary academic field that studies issues related to space exploration, settlement, and other issues from social-scientific perspective, and it is complementary to the so-called "hard" sciences. Collaboration on a much more extensive basis is already overdue.

Keywords: astrosociology, astrosocial phenomena, social sciences, humanities, arts, Mars, settlement

1. Introduction

Rockets and rocket scientists will be critical for getting humans to Mars, as will be those from the other STEM disciplines (i.e., the physical and natural sciences, technology, engineering, and mathematics). However, getting to Mars represents only part of the equation to help settlers to survive and thrive. Arriving safely is only part of the problem, and it importantly involves a great amount social-scientific input. On arrival, a safe physical habitat is essential, but a safe social environment within it is essential as well. It is important to remember that "expanding the human presence into space is about more than machines and missions. People are involved, with all their complexities and variations in beliefs, priorities, and behaviors" [1]. Social science matters on Earth and elsewhere [2]!

Thus, a major theme of this chapter is the emphasis on the two branches of science (i.e., the physical and natural sciences vs. the social and behavioral sciences, humanities, and arts). Specifically, this theme emphasizes the need for convergence between the two branches with regard to space issues [3]. For brevity from this point forward, whenever the term "social sciences" is presented alone, it also refers to all non-physical and non-natural sciences (as does the term "astrosociology"). This includes disciplines and fields such as sociology, anthropology, psychology, history, archeology, and all of the humanities and arts.

The arts are important as exemplified by the STEAM movement, which adds the arts to STEM, and the arts have always been a part of the astrosociology multidisciplinary academic structure. While the arts have received renewed attention, the social sciences and humanities are only recently beginning to receive their just due. Relatively speaking, however, they are still far behind the STEM disciplines in terms of recognized importance, as exemplified in the social media and literature. More about the arts follows in a subsequent section.

With this in mind, the purpose of this chapter is to explore the ways in which the social sciences can, and hopefully will, contribute to the human exploration, settlement, and thriving within the ecosystems on the Martian surface and the surrounding space environment. It is meant to complement the other chapters in this book. The purpose is not meant in any way to downplay the importance of the STEM disciplines and their contributions, but rather to focus on how the social sciences can supplement and enhance them. Separation has worked somewhat well in the past, but it is unlikely to continue its efficacy as humans begin migrating away from their home planet. The main theme here is that greater collaboration and cooperation is needed immediately, especially before mission planning is involved.

An important thing to remember is that this treatment of astrosociology on Mars deals with a future scenario in which a large and growing population is traveling to Mars. Smaller efforts are likely to precede this outcome. However, it is not too early to consider these issues now. In fact, many of the elements of discussion here will be relevant to the very first human mission to Mars.

2. Astrosociology defined

Before moving forward with this discussion, it is important to define the academic field of astrosociology. First, this field exists to encourage and foster greater participation of professional social scientists and humanists who may otherwise focus on other areas. Second, astrosociology exists to attract students with social-scientific orientations to study outer space issues, something that most of them would not have considered very long ago. Third, a very important aspect for gaining a greater understanding of space issues is to collaborate with space-based STEM fields; that is, the traditional space/aerospace community. These three components together make it possible for humankind to increase its scope of knowledge beyond just an overwhelming focus on one branch of science. The physical and natural sciences are enhanced by astrosociology because the human dimension complements them [4].

In 2004, there was very little progress among the social science disciplines, so it became a situation that needed resolution. Astrosociology is defined by this author, who also founded this academic field, as the study of astrosocial phenomena (i.e., the social, cultural, and behavioral patterns related to outer space) [5]. Astrosociology is not just sociology, as astrobiology is not just biology. It consists of sociology, anthropology, psychology, economics, political science, history, space law, and much more. The "sociology of outer space" or the "anthropology of outer space" can exist simultaneously with astrosociology [6], but this academic field is more inclusive and can work with the other approaches. The multidisciplinary aspect involves the social and behavioral sciences, humanities, and arts. It is important to keep in mind that astrosociology exists in order to balance the traditional approach so that the two branches (or cultures) of science can better collaborate in a more productive way. Astrosociology bridges the Great Divide between the two branches of science, as depicted in **Figure 1**. This clearly results in greater progress because the human dimension is added to the equation.

Figure 1.
The great divide bridged/photo credit: Jim Pass.

The purpose of astrosociology is to construct a cohesive literature and build a community of like-minded social and behavioral scientists, humanists, and artists who focus on astrosocial phenomena and strive to work together [7]. As mentioned earlier, there is a commitment to collaborate with those in the space-related STEM disciplines to forge a holistic understanding of space issues affecting those living and working on Earth and in space, including the interactive effects between the two. The overwhelming and detrimental fact is that astrosociologists remain far behind physical and natural scientists in terms of contributing their equal share of input to the outer space knowledge base. Relatedly, there is much historical and behavioral research from the social sciences and humanities that is applicable to helping plan for social life on Mars.

3. Physical science and social science contributions

Accordingly, this author founded astrosociology in 2004 due to the lack of social-scientific input into the study of astrosocial phenomena [8]. This was always a relative situation, however. When it is compared to the input by physical and natural scientists, the overwhelming emphasis remains on the STEM-related issues and concerns, and therefore the overwhelming funding goes to non-social-science individuals as well as private and public entities. The equal distribution and attention to social science will likely never reach that of physical science, but a substantial increase is sorely needed at a time when sending hundreds of people to Mars is seriously being considered by the Space Exploration Technologies Corporation (SpaceX).

To be fair, scientists and scholars from the physical and natural sciences such as the search for extraterrestrial intelligence (SETI), astrobiology, astronomy, planetary science, planetary defense, space architecture, and engineering have recognized the potential of the social sciences and humanities, and they have worked with social scientists in some instances. Examples include Ben R. Finney [9], Christopher McKay [10], Penny Boston [11], and Seth Shostak [12]. Too many exist to reference here so the reader is urged to do research to find them. An early place to start is this reference from 1988, which is an edited book that actually includes physical *and* social scientists [13]. This additional publication is a good example of a social scientist (an anthropologist) and a space-related physical scientist (an astronomer) working together [14].

This recognition is important because it opens the door to the realization that the social sciences are relevant. It is also problematic because physical and natural scientists are not well versed in the social sciences unless they work with social scientists. This area of knowledge, which includes sociology that is over 200 years old in terrestrial societies, is more relevant to study social science issues related to space—that is, astrosocial phenomena—because they possess the expertise. For this reason, the best overall approach is for the two branches of science to cooperate and collaborate on a formal basis.

Social scientists and humanities scholars have, in fact, contributed to the study of space issues. This trend increased somewhat during the early days of the space age when science did not know if astronauts could functional well mentally and physically in space environments [15]. Rudimentary space psychology existed to monitor National Aeronautics and Space Administration (NASA) astronauts before the United States launched its astronauts. It continued starting with the Mercury program during training, missions, and afterwards. Space psychology and psychosocial research continues today [16, 17]. Space historians also played a large part throughout the space age and they also continue to do so into the "NewSpace" age, which is characterized by commercial space activities [18–21]. Many other social scientists have also contributed in meaningful and less impactful ways. Again, there are too many examples to go into detail here, but they are listed in the reference section as a good starting point [22–37]. There are many more examples, and even this listing may seem quite overwhelming, but these social scientists represent just a trickle of contributions compared with the physical and natural sciences during the advent of the space age through today. There was a concern about the lack of social science input. For example, Rudoff voiced it succinctly in 1996 when he asked, "And where is sociology?" [25], p. 75. The other disciplines were also silent.

"The social sciences and humanities cannot afford to remain silent, and the traditional space community cannot afford to allow them to remain so" [4]. One of the most frustrating realities is that many social scientists work in isolation, which is a central reason why this author created astrosociology. It is intended to provide for a community in which social scientists can more easily collaborate and create a cohesive and easy to locate literature. This is changing, however. It is hoped that the creation of astrosociology and the work of the Astrosociology Research Institute has played a part although the increasing impact of astrosocial phenomena in larger culture have also contributed apart from it as well. It is hoped that the future will yield a much better level of collaboration between the two branches of science as time passes.

4. The Astrosociology Research Institute (ARI)

In 2008, the Astrosociology Research Institute was created by this author in order to advance the development of astrosociology beyond 2004 when this author first founded this academic field. The mission of ARI is to facilitate opportunities for others and to contribute directly to the development of astrosociology. ARI is a 501(c)(3) nonprofit educational and research organization that depends on donations to advance its mission.[1] It also depends on the participation of volunteers

[1] ARI's webpage is the following: www.astrosociology.org, which includes the Virtual Library and donation page.

Figure 2.
The ARI logo and the planet Mars.

to contribute to its development. Regardless of its resources, ARI continues to promote and advance the need for the social sciences to become more involved in research and educational efforts related to outer space exploration and related issues.[2] Settling on Mars requires more research by professional social scientists and their students, as **Figure 2** represents. At the moment, too few opportunities exist for social science majors and graduate students to pursue astrosociology at their universities and colleges.

Therefore, another important part of ARI's mission, besides encouraging social science and humanities professionals to take part in astrosociological education and research, is to help interested students to gain access to the study of astrosociological issues. Related to this is the "Astrosociology in the Classroom" program. It encourages educators and programs to include astrosociology in their classes and to even create entire courses.[3] ARI is dedicated, among many other space-related issues, to getting many more social scientists involved with Mars education and research focusing on a number of fronts, many of which are covered in this chapter.

Students pursuing STEM-related space science degrees are encouraged to also take social science, humanities, and art classes. Professionals from the space sciences seem to be increasingly recognizing that the so-called "liberal arts" are valuable to their own research. This author is also being contacted much more frequently recently about astrosociology. A more holistic education results in a more holistic approach to understanding astrosocial phenomena.

---

[2] Materials from this author and others, which include issues of the *Astrosociological Insights* newsletter and *The Journal of Astrosociology* are available at no cost in the Virtual Library at www.astrosociology.org.

[3] The first ongoing course was introduced at Harvard University by Dr. Gerhard Sonnert who is now a member of ARI's Advisory Board and Editorial Board for The Journal of Astrosociology.

5. The meaning of "Astrosociology on Mars"

"Astrosociology on Mars" reflects the need to place social scientists—astrosociologists—on the Martian surface along with the expected STEM-oriented individuals. Both branches of science, the physical and the social, provide benefits to settlers, as they must cope with issues in their physical and social environments. They must interact within their physical space with one another. "Astrosociology on Mars" most importantly refers to the need to put *astrosociologists* on Mars!

More broadly, the "Astrosociology on Mars" concept refers to the vital need to include social and behavioral scientists, humanities scholars, and space architectural artists in the planning for human-based ecosystems in the Martian environment on a permanent basis. Just as importantly, it refers to the vital need to include them in these ecosystems in order to help set up social institutions and continue to study the behavioral patterns of the members of social systems as their populations increase: from bases, to communities, to space societies. Each significant population increase results in exponential rises in complexity in the population's behavioral patterns and thus an increase potential social problems.

The social sciences are vital to the functioning of terrestrial societies and there is predictably even a greater need for this type of research in Martian settlements. Unlike isolated habitats on Earth, such as underwater facilities, submarines, and Arctic and Antarctic bases, the perception of returning back to terrestrial civilization seems unlikely, or at least months away, which can result is mental and social problems for some individuals. Without the input of astrosociologists on Mars, chaos and conflict in addition to other forms of deviant behavior is much more likely to occur, as the stressors generated by confinement and isolation on another planet can cause great behavioral disruptions. Research on Earth focusing on similar settings, both related to space and unrelated directly, have provided invaluable knowledge and insights. See these publications, which provide important examples (although a large number of others exist as well) [13, 38–41].

Astrosociology on Mars is essential to humans settling there because their lives depend on social order and cooperative interactions. Thus, they are just as important as the life-giving contributions of the physical and natural sciences and engineering technologies. The physical structures must make staying alive possible, so they are necessary although they are not sufficient, and thus the social structures must also exist to make social life possible. The social sciences and physical sciences to work together, as both are necessary for life at the planet Mars. They must work in concert with one another. As such, astrosociology belongs in the Martian settlements along with the physical sciences. Generally, astrosociology exists to bridge the chasm between the two major branches of science as depicted in **Figure 1**.

6. The astrosociological imagination

The concept of the sociological imagination was introduced by sociologist C. Wright Mills in 1959. In this sociological context, he stated that it is important to recognize and understand relationship between the self and society, to possess "the vivid awareness of the relationship between personal experience and the wider society [42], p. 6." Specifically, he stated that "The sociological imagination enables us to grasp history and biography and the relations between the two within society. That is its task and its promise [42], p. 6."

Application of the sociological imagination to the field of astrosociology allows for the introduction of the *astrosociological imagination*. "The astrosociological imagination is a high-level conceptualization; it is an insightful way at looking at the world that allows a person possessing and exercising it to make connections between his or her personal world of experiences and the macro-level (larger scale) existence of astrosocial forces [43]." It is an understanding that personal experiences alone do not reflect the totality of reality.

Thus, the specific focus on the vivid awareness of the relationship between personal experience and the impact of astrosocial phenomena is an important case because a substantial portion of any given society fail to recognize their relationships to the overwhelming and increasing influences of space activities. This includes policies and actions. It also includes the hidden forces that tend to exist out of the attention of the average citizen. Many citizens regard the money spent to explore space and humankind's place in our universe as a waste of tax dollar and private investment as a waste as long as terrestrial social problems require attention. This Earth-centric view lacks the vision necessary to understand the present impact of space activities and the foresight to imagine how much more they will impact on humans wherever they reside.

Why is the astrosociological imagination important to humans settling Mars? A key reason is that the astrosociological imagination broadly speaking is vital to recognize the relationship between individuals and the larger Martian social society. More important in many ways is the fact that living on Mars means that space has a much greater impact on everyday social life. Therefore, possessing a strong astrosociological imagination becomes vital to not only recognizing less obvious impacts on people's lives, but also vital to survival. The physical environments, and thus the ecosystem within the habitat are much more dangerous and require acute attention to the obvious social forces as well as the underlying patterns. Lacking the astrosociological imagination on Earth is disappointing, but lacking it on Mars can become injurious or even fatal.

7. Planning before launch, during transit, and once arriving on Mars

7.1 On Earth, before launch

On Earth, planning must involve the social construction of the settlement as a theoretical model that takes into account the intricacies of a social system that has never existed. Depending on those who lead the planning effort, the type of Martian society will differ so that its characteristics and functional details align with their priorities. Planning must occur on Earth before the first launch so that the earliest days of settling the Red Planet becomes as optimal as possible. Planning an ecosystem early in the process increases the odds of success [44].

Astrosociology on Mars begins as "Astrosociology on Earth" in the sense that the human sciences must be involved in providing invaluable insights as to the astrosocial phenomena that will inevitably impact on the success of any space mission, whether temporary or permanent in nature. The contributions by social scientists from the planning stage onward remains undervalued to a perplexing extent, but their importance needs to be understood in terms of their true impact. Too few contributions can result in a failed or chaotic outcome while an adequate level of contributions can assist with the physical aspects and, more directly, on the behavioral aspects while in transit and after arriving on the Martian surface. Currently, the social sciences and humanities continue as a neglected category in the planning stages and beyond.

Sending tens, hundreds, or even thousands of humans to Mars in a relatively short period of time is not recommended without input from astrosociologists. As a theoretical enterprise, consider what would occur if 100 individuals were sent to a deserted island without the knowledge regarding how to construct a functional social system. The trial and error attempts to reach a civil community will inevitably result in harm to some individuals without social-scientific expertise being infused in the planning.

Just as the construction, planning, and testing of the hardware is an essential component, which gets people to the red planet, construction of the social system on the ground is also vital. While it may take 6 to 8 months to travel to Mars, life in the Martian settlement will last lifetimes and perhaps generations if all goes well. It is therefore imprudent to expect settlers to construct the social system of a functioning settlement only after they arrive.

7.2 Life in the spacecraft during transit

During life in the spacecraft, the theoretical construction of the settlement must become implemented as a practical, functioning version. For example, the culture must be understood by the population and rules of behavior, the social norms, must be accepted, including penalties for violations. "A central assumption made here is that the population that ultimately leaves our planet together must be socialized into a single social order that exists before boarding their spacecraft; and they must largely accept the ideas (including values) and the norms or social rules that protect these values that comprise a single culture, which is part of the social order [44], p. 3." Details about culture and social structures follow in subsequent sections.

The long journey to Mars makes it imperative to provide structure and understanding among those in the population about what types of behaviors are acceptable and which are not. This stage of settling Mars provides a shakedown of sorts that put the plans made on Earth to the test. It must be structured in order to acclimate travelers to the social system and culture, and so that they know what is expected when they arrive at Mars.

This stage of the settlement mission will be characterized by boredom, largely because a great majority of the voyage will be automated [45]. There will be extreme challenges on the long-duration trip to Mars [46]. The best way to meet the trials involved so that stressors are minimized is for the physical and social elements to work together. The astrosociological approach involves collaboration and taking advantage of social-scientific knowledge available today and that learned in the future before the mission begins. The more knowledge that can be implemented in the planning, the better the experience on the voyage to Mars.

7.3 Once arriving at Mars

Once arriving at Mars, settlers must put the predetermined plan into operation. The transit phase will allow for adjustments to the original plan to be put into practice, as problems and other unanticipated issues will have undoubtedly arisen. With a good original plan as a starting point coupled with the lessons learned during the trip to Mars, the construction of the social system and the physical components of the habitat can help ensure a better organized starting point, even with the difficulties that will exist. This eventually would not be nearly as practical without the original plan that was formulated on Earth, and of course, the great bulk of this chapter focuses on this arrival stage.

8. Physical and social environments

Physical environments include the spacecraft and habitat structures while the social environments include the interiors of these physical environments where humans live and interact. It is important to distinguish between the two. Planning regarding the physical and social aspects must receive equal attention because the construction of the social ecosystems is just as important for the survival of the inhabitants as the construction of the physical components [47]. "Although engineering solutions can construct a physical environment that can sustain the population on a biological basis, this capability cannot ensure the success of the settlement due to its inattention to the critical issues related to the social environment" [48], p. 554. They must be also integrated as one whole entity. If one of them goes "out of whack," it places strains on the other. There is a balancing act in which living in a Martian habitat constantly requires monitoring.

Determining where to place a settlement, such as choosing between the surface or lava tubes, for example, can have implications down the line for the social environment. "Lava tubes on the Moon and Mars are of interest for space settlement because caves have been proposed as natural shelters that future human explorers could occupy. Caves would in principle protect dwellers from surface radiation, wide temperature swings, micrometeorite impacts, and rocket exhaust blast" among other benefits [49]. Another potential benefit is that life support requirements could be simplified. This could thereby reduce the rate of resource use due to fewer precautions that would be needed to protect settlers and their equipment.

Alternatively, a surface settlement would need to implement greater resources and strategies to protect the settlement and its occupants. Existing on the surface could provide easier access to the objectives of settlers such as accessing water, greater mobility to different areas including the use of rovers, and better access to searching for Martian life. The decision regarding which type of environment is chosen will be based on a cost-benefit analysis based on the objectives of the leadership. Another alternative, if practical, is taking advantage of both options in a fairly small area. Again, physical and social criteria must both be part of these calculations.

As this chapter emphasizes, the social environment cannot receive too little attention by planners during the first two stages and astrosociologists must be ready to conduct their research once construction of the Martian settlement begins. Future Martians will need to cope with the same interpersonal issues that populations on Earth do, but in far smaller and more confined social settings. And, in fact, more challenging living conditions can easily aggravate social interactions due to the harsh Martian conditions outside of the habitat components including the mainly carbon dioxide nonbreathable and thin atmosphere (see **Figure 3**), cold outside temperatures, solar radiation, seasonal dust storms, and the lack of even the simplest conveniences.

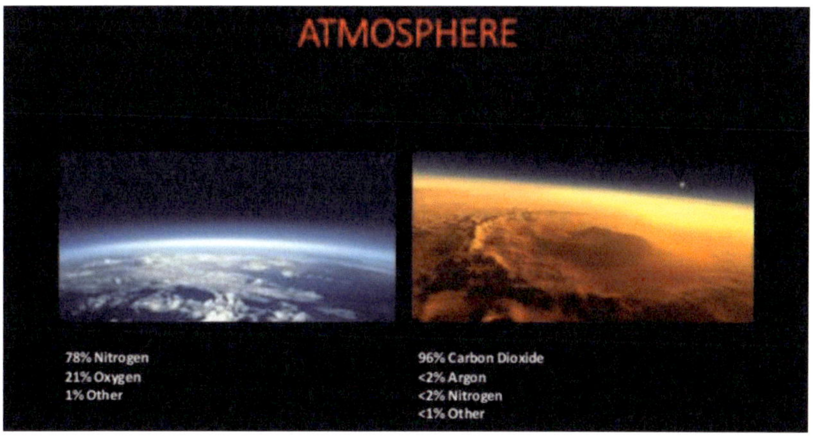

Figure 3.
The comparative atmospheres of Earth and Mars/photo credit: NASA.

Habitat structures must provide adequate opportunities for Mars settlement citizens to be able to socialize, to interact when they need to be around other people. Compartmentalization, which is often depicted in artistic renderings, is not a prudent design when carried too far. Privacy is important although isolation is not healthy when socializing becomes difficult under these types of isolated and enclosed conditions. The social environment, the ecosystem inside the habitat, must become increasingly designed as it grows in size to limit isolation among those in the growing population. This should be integrated into the original plan. Space architecture should also make the design within the physical structures more like a community and less like a spacecraft or base.

9. Space architecture and art

There has been an increasing need to develop the field of space architecture for designing and building progressively complex spacecraft, especially those carrying humans, and for getting a jump on the habitats slated for the Moon and Mars in the probable near future [50]. "Space architecture is the theory and practice of designing and building environments for humans in outer space [51], p. 890." Mars habitat simulations conducted by the Mars Society [52], NASA, ESA, and others have tested hardware and have increasingly focused on social-scientific areas of research as well. **Figure 4** shows the Mars Society's Flashline Mars Arctic Research Station located on Devon Island, Canada. Work in these types of analogs produce valuable science that amend submarine, aircraft carrier, and terrestrial Arctic station research efforts among others.

Living in a space habitat for the rest of one's life presents adjustments, which impact different individuals in various ways. The transition from the long trip to Mars aboard a spacecraft characterized by weightlessness to one that characterized by a gravity field that is one-third of Earth's will present settlers with physical and psychological adjustments that must be made. "Through the body, sensorial data and emotional response interact to create symbolic meaning that ultimately impacts the development of new spatial habitats. The creation of such 'places' requires the understanding of the human environment interface and integration of territories

Figure 4.
Flashline Mars Arctic Research Station/photo credit: Mars Society.

that range the psychological, social, ergonomic, anthropologic, perceptual, and anthropomorphic that radiate into interconnected and intra-disciplinary fields [53]." Astrosociologists on Mars also entails conducting research while on the voyage to Mars. Understanding how well or how poorly various passengers faired in weightlessness presents the researcher with challenges that entail helping each individual adjust to his or her new life on Mars.

Functional hardware alone does not result in a functional society. The human dimension, which involves social interactions and reactions to living conditions, is just as important on a medium and long-term basis. At its most basic level, the construction of the physical habitats and other structures will be built with operational functionality in mind for various purposes without regard to the humans that inhabit and work within them. Moving beyond that standard takes into account the human beings and the ecosystem in which they live. The physical structures have direct and indirect impacts on the social structures in the various habitat components.

Space architecture, when implemented to benefit the people, moves beyond the minimum standards of function. It adds layers onto the basic survival standard. Space architecture can make living in a Martian settlement survivable, which is, of course, vital for any life, but it can also make the existence there livable and even enjoyable. The adventurous attitudes present in the initial planning and training on Earth will likely give rise to practical thoughts about the hardships that lay ahead during the voyage to Mars. Therefore, the architecture of the spacecraft is important, but the construction and details paid to the social environment of the habitat components is even more critical because that is where they will spend most of their time, even their entire lifetimes.

Art, as part of the construction process, can also provide an esthetic that reminds inhabitants of Earth. While "space art" generally depicts the wonders of universe, Earth-based art can go a long way to improve living conditions on a psychological level. It can also provide a sense of homeness as opposed to a stark minimalist enclosure. A military style is probably acceptable for small and temporary quarters such as the International Space Station. A military or quasi-military purpose is the legacy of the space age thus far, but individuals with nonmilitary statuses and roles will not adjust well in that type of political system.

A large and growing population will need to establish an organic form of solidarity, which is social cohesion that is based on members of the population forming a dependence on one another, which includes diverse statuses or social positions [54]. Contrastingly, the mechanical form of solidarity is characterized by most of the people sharing common values and beliefs, which is workable for a small crew but not feasible for a growing settlement [54]. Astronauts at NASA are trained very intensively and similarly in order to perform extremely well-choreographed scheduled tasks. They are taught to improvise during emergencies though these are rare occurrences. In contrast, settlers on Mars will be much more diverse with educational backgrounds and experiences that may have little or nothing to do with running the settlement politically, engineering-wise, or performing various forms of maintenance.

Architecturally, the best approach is to make social settings within the habitat components as reminiscent of Earth as possible rather than simply following a functional scheme that fails to take advantage of the psychological, social, and cultural requirements of settlers that transcend meeting engineering standards. This will be most important for early habitats. As habitat structures increase in size and sophistication, the needs of the humans on Mars should become higher priorities. At first, in contrast, the engineering issues will prevail in order to allow for settlers to survive on the Red Planet.

10. Martian settlement cultures

Every base, community, city, or society on Mars—or anywhere else—will include a culture that shapes behavior. It defines how we live in a particular society and includes the characteristics of the people who live in that society. The popular meaning of culture that includes popular movies and music, as two examples, does not apply to this discussion. The social-scientific definition, which is discussed in this section, is quite different.

Settling Mars is a cultural invention. "Of course, all rationales for space activities are cultural [36], p. 31," which means that how settlers organize their physical and social constructions will be based on ideas carried with them from Earth. "Culture is inescapable [36], p. 31". In the case of settling on Mars, rationales include exploration in order to gain scientific knowledge, the adventure related to living on another planet, and ensuring that the human race survive in case of a global catastrophe on Earth. Mars has long fascinated humans and has gone through a number of different phases, the last of which is viewing Mars as the new frontier, but this cultural idea could well result in transposing past terrestrial problems onto the Martian surface so that not enough would be learned from history [55]. Bringing imperialism and various social problems to Mars is something that needs avoiding whenever possible. Harmful cultural ideas represent unneeded baggage better left behind.

Here, the definition of culture encroaches on social life much more than the popular meaning even though popular notions can represent reflections of the social-scientific definition. Interestingly, however, many of the impacts of culture are not always recognized and understood by the general public as forces that affect them. The analogy has been made by many that "culture is to humans as water is to fish." This analogy points out that fish depend on water to survive just as humans depend on culture to survive in their society, but neither recognizes it. At Mars, the evolving culture will have the same impact on settlements and settlers. And, as discussed earlier, a well-articulated culture should be part of the planning process before lifting off for Mars.

In 1970, sociologist Robert Bierstedt identified three dimensions of culture as consisting of ideas (including values), norms (i.e., social rules), and material culture [56]. The ideas in culture provide citizens with the important priorities and acceptable standards of conduct. Norms exist to protect values. Material culture consists of the physical creations of humans in society (i.e., the physical habitat components, rovers, and spacesuits), which will become increasingly important as they are constructed and their meanings become integrated into the larger culture. These meaning can evolve into changed ideas over time.

While important, the larger culture does not totally affect all individuals and social groups in the same way. There are contrary forces in any society. Settlers from different terrestrial cultures can interact in ways that result in conflict within the same settlement or between different settlements, as can religion-based differences, for example. Subcultures exist that may oppose some aspects of the larger culture while countercultures oppose the dominant culture itself. Thus, while subcultures may produce social movements, countercultures can produce revolutions, even in a Martian settlement. Astrosociologists on Mars will need to monitor contrary patterns of behavior that will inevitably arise.

11. Social institutions on Mars

Social institutions regulate behavior, as they are part of the social order of a society. They consist of people who share a common purpose. This definition is at odds with the common public view that may view the courts or schools as institutions because they are components rather than the entire institution. An important consideration is the fact that social relationships become progressively complex as the population increases in size. As such, social institutions become increasingly more important and their missions become increasingly complex and thus more difficult.

Any settlement on Mars must involve well-defined social institutions in order to control behavior. Haphazard plans of constructing the settlement both physically and socially will result in a level of chaos that can easily lead to the failure of the settlement to sustain itself even with adequate supplies from Earth. Moreover, the physical and social constructions are intimately interconnected, which means that addressing both types as a single system is vital. As such, planning before the mission begins, including devising a culture and a set of norms, cannot be overestimated even while most existing plans seem to downplay or even ignore this requirement [44].

Below are brief issues related to some of the most important institutions of any human social system. Each requires substantially more attention in the future. Nevertheless, pointing them out here is a good start, as they are brought to the forefront of a Mars-based discussion, which does not occur nearly enough.

11.1 The economy

The economy is an important institution to consider because, among other things, it determines how resources are distributed to other institutions, groups and other entities, and individuals. At Mars, the economy initially will be rather straightforward and dependent on supplies from Earth. As time goes on, however, Martian resources may well build production in several areas that can lead to exchange functions with entities on Earth. In general, how fair the resources are distributed is an important issue for astrosociologists.

Settlers must feed themselves, which means that they will rely on sources of nourishment. This will become increasingly more difficult as the population increases. While growing food in greenhouses will be necessary, most resources will come from Earth. The early Martian economy will be difficult to sustain requiring much hardship at first. Over time, the economy can begin to flourish as settlers take advantage of in situ resources and develop unique services such as tourist destinations such as rover and lander monuments, or hotel lodgings.

11.2 The political system

The political system is a vital component that determines how the settlement functions. Who governs? Will the leadership be characterized as a military, quasi-military, democratic, authoritarian, totalitarian, or charismatic political system? The structure and actions of politics can shape the structure of the social system and quality of life of citizens. Therefore, the comparison between two settlements could very well reveal significant differences between the two on a number of social and cultural dimensions. The experiences of each of the populations could be quite different. What type of class system will exist upon landing, or even en route, and how will it evolve in the future? How will different categories of people be treated? Will discrimination be an overwhelmingly harmful characteristic?

As we have witnessed on Earth, the leaders change over time and they can move in extremes such as from a democracy to a charismatic or even totalitarian orientation. Astrosociologists will need to conduct research to determine how the political system functions and what changes occur. This is another example of the importance of social scientists on Mars.

11.3 Criminal/juvenile justice system

Criminal/juvenile justice system operations exist to ensure social order by acting proactively and reacting to unlawful behavior. One thing that is certain in all human societies is the fact that deviance is a cultural universal and will exist in any social system, including on Mars. Therefore, it is important to devise a preconception of a justice system that can handle violations of the law and conflict that may arise even during travel phase before reaching Mars. Relatedly, laws must be understood by the population and protected by law enforcement. Equality in policing and court decision making must be seen as fair. Otherwise, protests can arise as is evident in terrestrial societies.

Enforcement of other types of norm violations such as those involving the health and safety of settlers is also an important consideration, especially in an enclosed and confined living situation in which a single mistake or act of sabotage could have deadly consequences [57]. Overwork could lead a person to cut corners during maintenance activities or inspections. Unsafe conditions may also result in protests and more serious reactions. This is an important aspect of life in a Martian settlement that requires constant maintenance and inspections in order to ensure health and safety standards for the protection of the population.

11.4 The family

The family is an institution that receives little attention in the midst of attention to other issues related to settling on Mars. While children would be highly unlikely to go to Mars initially, couples will very likely be in the population. Additionally, romantic relationships will undoubtedly develop and that could result in children down the line. Support for families would be required including their rights and supportive policies.

Family structures will differ, as they do here on Earth, which is something that is important to monitor. Some structures may receive condemnation by official sources including other institutions that may result in inequality suffered by family members. Hardships caused by the harsh conditions on Mars from the environment outside and the ecosystem inside (including isolation) can produce stressors and strains between family members and couples that endanger their relationships even when inequality is not the root problem.

11.5 Religion

Religion can foster fellowship and community within the population, but it can also cause conflict between religious groups and individuals with different religious backgrounds. Atheists will also impact on this social dimension. Some forms of tolerance policies should exist so that harmony rather than conflict exemplifies the religious dimension. In any case, "living in close quarters in hazardous environments will make interreligious dialogue all the more important..." [58]. Settlers will need to find a way to get along if the settlement will be characterized by a minimum of religious-based deviance.

Politics can influence how the institution of religion interacts with non-religious institutions. Will there be a sort of separation between religion and state? Conversely, will the settlement's government promote secular society or be based on a religious dogma such as Christianity or a sect of the Muslim faith? This example demonstrates how social institutions interact and complicate social life in any society.

There is another interesting possibility. Living on Mars could well involve social forces that produce one or more new religious groups in the form of cults (new religions), sects (fully branched off religious groups), and new denominations (slightly offended branches off religious groups from their established churches). These types of evolutionary occurrences could add new complexities to the social system. Cults and sects may challenge other religious groups or the Martian government itself. Religion is an inevitable dimension of social life that requires social-scientific scrutiny, just as all others.

11.6 Education

Education is important, not only for children and young adults who would arrive later in the migratory process, but also for older adults who must learn how to live in the new settlement. New members need to learn the differences from Earth that they first encounter in a substantially new and different ecosystem. For example, they cannot easily go outside of the habitat components and must learn to live in a confined and enclosed ecosystem.

Additionally, settlers, especially early in the construction of the settlement, will need to learn new skills and concepts. They will need to conduct multiple tasks and teachers will need to pass them on to new settlers as they arrive. The needs of the settlement will change over time so that arrivals will not always possess the most desirable attributes.

11.7 Health care

Health care is an extremely important issue in harsh environments such as that of Mars. The potential for illness absolutely increases due to the harsh Martian conditions that include solar radiation and the one-third gravity field. Ethical issues include right-to-die disagreements and the potential inequitable distribution of medical resources, which will be limited [59]. Conflicts or disagreements between patients and/or their family members and medical practitioners can result in conflict. Social problems such as racism, sexism, and classism can affect the quality and very implementation of life saving interventions as witnessed on Earth. (See the medical astrosociology section below for additional details).

11.8 The military

The military on Mars may exist in order to protect one settlement from another's transgressions once they begin to appear. Another possibility is providing defense against entities on Earth that they fear may interfere with their operations and priorities. This latter situation could be complicated if they a Martian settlement possesses valuable Martian in situ resources that an Earth-based entity covets. A military presence may exist just because it is how things work on Earth. The possibility of a military existence is something for astrosociologists to monitor because its members may interfere with civilian authority in inappropriate ways.

11.9 Sports and the media

Sports and the media must not be viewed as extraneous to a well-functioning Martian settlement. As witnessed in terrestrial societies, the absence of sports and simultaneously the importance of entertainment and news in their various formats demonstrate that these two social institutions provide important lessons for settling beyond Earth. It is important for the citizens of a Martian settlement to distract their attentions away somewhat from their hardships associated with living on an isolated and potentially deadly planet. Two forms of sports and media consumption exist.

First, on Mars, planners of the habitat modules—and to the extent possible aboard the spacecraft—should focus on establishing community areas because sports and media outlets together represent extremely relevant outlets for the well-being of individuals. Live sports and entertainment events such as concerts and plays in the settlement can contribute to a rise in psychological well-being and group cohesion. Recreational activities are also important. This is different from interacting with coworkers, and thus, off-work activities must supplement it. A Martian Internet should be part of the planning process before launch. Encouraging a focus on the benefits of living on Mars, including the social interactions with others living there, is important to mitigate feelings of isolation and homesickness.

Second, the consumption of sports and media content from Earth can help negate any psychological hardships that are exacerbated by communication delays with those on Earth, as they will provide various levels of frustration among settlers. Connecting to Earth-based cultural trends, events, and people is vital, especially for new arrivals. Therefore, recorded sports events, entertainment programming such as television shows and movies, and news recordings will mitigate some of that despondence. Recorded messages that can be played without the prospect of interaction can help although it also emphasizes the disconnection among individuals on each planet. Recorded content cannot replace communicating interactively with loved ones, friends, colleagues, and others; however, a combination of communication formats will be required. Finally, a warning of sorts is important. Sports and the media must not receive undue attention because, together, they can act as a pacifier that glosses over negative sociocultural realities such as social problems that could exist including discrimination or the support of other harmful social conditions. The prospect of placation of the citizenry will depend also on the agendas of the other social institutions including the political influences. Precautions to avoid instances of community harm will require ongoing oversight by social scientists in settlements away from Earth.

12. Social problems in Mars settlements

Social problems can harm most people or specific categories of people, and they will exist in any Martian settlement. Many of the problems faced within, and between, terrestrial nations are difficult to address and mitigation is often not a priority among political leaders. Given the seriousness and ingrained nature of social problems, is it a good idea to spread these harmful behavioral patterns into our solar system and beyond?

On the one hand, an argument exists against migrating beyond Earth's atmosphere and staying put on humankind's home planet [60–62]. This position disfavors settling Mars based on a crisis of faith in which people cannot be trusted to set their differences aside and create a society that lacks social problems. Thus, space exploration, including the settlement of Mars, is not a guaranteed outcome, or at least it is not necessarily easy to put into motion and sustain. There are forces that prefer to slow down sending humans to Mars. There are other forces that want to limit the number of humans who go there in order to reduce contamination levels and the number of contaminated sites.

The contrary argument is that we *must* settle Mars despite humankind's history of conflict, inequality, and production of social problems. Advocates favoring settling other planets and space environments argue that social problems can be minimized in terms of their negative impacts based on the perceived ability to learn from history so as not to repeat humankind's worst errors. The movement to going to Mars seems inevitable with SpaceX and other private companies gearing up to send people beyond the cislunar space environment. This author foresees the inevitability of settling elsewhere in our solar system *and* the manifestation of serious social problems occurring together just as the existence of social problems did not stop exploration on Earth.

With this in mind, it seems prudent to keep the argument against migrating into our solar system in mind while focusing on how best to make a settlement function as fairly and equitably as possible. It is best to include both types of advocates in the planning process in order to be exposed to voices that point out discriminating strategies and other forms of harm to potential settlers. An overwhelming focus on the hardware that takes attention away from the human dimension, which astrosociologists focus upon, will result in skimping on things that may bring discomfort or even danger to human beings. "Astrosociology on Mars" refers to the beginnings of the mission on Earth through establishing the settlement, and beyond, as human interaction and the functioning of the physical aspects of the settlement continue.

Social problems will continue in various forms in Martian settlements [63]. They will differ from one settlement to another just as social problems differ among various nations on Earth. Part of the determination, at least for the start of a settlement's beginning stages, will be determined by who funds the mission and the composition of the leadership. The nature of the settlement will also bear on how it functions and the priorities. A scientific mission will differ from a communal type of orientation, for example. Another settlement may be characterized by a mixture of priorities as well.

Settlers will need to deal with the various types of social problems, which include injustices. This is because they will bring them along due to being socialized in terrestrial societies. They may well possess attitudes that lead to unwanted behaviors. Inequality will undoubtedly exist that includes sexism, racism, poverty/social classes, religious conflict, and political dissent. How well the socialization during the planning and traveling stages could serve to curtail the impact of these problems is important although they will exist to some degree and must therefore receive serious attention. Astrosociologists on Mars can help with this by identifying problems and proposing mitigation strategies.

13. Space medicine and medical astrosociology

13.1 The healthcare system

The healthcare system is an extremely important social institution briefly described in an earlier subsection. Relatedly, space medicine is vital for the health and safety of settlers on Mars. Traditionally, space medicine has focused on the effects of spaceflight on the human body, which strongly involves the biological and life support issues. It is defined more broadly as the "practice of all aspects of preventive medicine including screening, health care delivery, and maintaining human performance in the extreme environment of space preserving the long-term health of space travelers [64]." Thus, preventive and reactionary efforts are involved, as countermeasures are not always effective.

These types of definitions apply mainly to living in a spacecraft and they focus on human factors rather than social-scientific principles. Human factors and ergonomics are important although not all practitioners are also social scientists. The two are complementary, however. This is important during the voyage to Mars though social science should apply more strongly even in weightlessness. However, when the focus extends to the settlement where the gravity field changes from zero gravity to one-third of Earth's, Martian medical practice will change in order to adapt to yet unknown biological effects. It will depend on the specific conditions of the planetary environment that will differ from weightlessness in outer space. Intensive research will be necessary at that point.

Before that point, the trip to Mars will likely involve a considerable time spent in near zero gravity (although artificial gravity is only a slight possibility because this option adds additional cost and difficulties in the solving engineering issues for the spacecraft). This means that the same problems faced by astronauts within the cislunar environment will be experienced by the settlers during the voyage although it may well be worse due to going from microgravity to zero gravity unless preventive actions are taken. It will be vital to keep each individual as physically fit and healthy as possible in order to reduce the physiological problems that could exist upon landing.

Some of the well-documented health issues that affect humans during space travel include radiation poisoning, DNA damage, cardiovascular stress, bone density loss, swelling of the optic nerve head, dehydration, cognitive decline, increased risk of mutations and cancer, and various molecular, behavioral, and physiological changes compared to those who continue to live on Earth. These types of detrimental effects were confirmed as a result of the NASA twin study involving Scott Kelly who stayed aboard the International Space Station (ISS) and Mark Kelly, also an astronaut, who remained on Earth [65] (see **Figure 5**). Living and working on the ISS for up to 6 months presents measurable health risks and problems, but this is mild compared to what can happen during a voyage to the fourth planet.

13.2 Medical astrosociology

By contrast, though also complementary, to space medicine, **medical astrosociology** is the study of the social, cultural, and behavioral patterns (i.e., astrosocial phenomena) that affect medical issues in space ecosystems such as settlements on Mars [66]. The reason this author created this subfield of astrosociology is an attempt to attract social scientists, especially those who have studied Earth-based medical sociology and medical anthropology in addition to students who may be interested in studying medical astrosociology. The major focus is on the sociocultural and psychological forces that affect space travelers and settlers. The effort is to bring in more social science professionals and students to work with those in space medicine.

Medical astrosociology has become part of the trend that has been moving beyond a concentration on only biomedical concerns so as to include social-scientific dimensions of health. NASA's human research program, for example, "includes many facets

Figure 5.
Astronauts Scott and Mark Kelly/photo credit: NASA.

of human space travel such as environmental factors, exercise physiology, habitability, human factors, medical capabilities, physiology, psychosocial and behavioral health, and space radiation [67].” An important aspect is determining how traveling in outer space and living in various space environments affect human biology. Producing countermeasures to harmful effects will represent an ongoing set of challenges.

13.3 Behavioral health

Behavioral health is an important aspect of assessing the well-being of a person in an isolated and confined physical environment, which involves both space medicine and medical astrosociology. It can be defined as “a lack of neuropsychic dysfunction, and the presence of high levels of personal adjustment, cordial interpersonal relations, and positive interactions with the physical and social environments [51], p. 890.” When considering behavioral health, countermeasures need to be identified arising from human, environmental, and external factors [68]. The overall wellbeing of individuals must be evaluated as a combination of several medical and psychological problems along with social and cultural forces that impinge on the individual and the ability to interact productively with others. For Mars, these types of forces remain unknown until humans actually set foot on Mars. It is a holistic approach.

13.4 Space medicine and medical astrosociology

Space medicine and medical astrosociology are, in fact, complementary approaches, then, that focus on the health and safety of spacefarers and space settlers, and thus even greater convergence is needed. Together, they provide a comprehensive assessment of the condition of the patient in terms of biomedical impairment on the one hand and sociocultural, ethical, and psychological condition on the other hand. Both approaches focus on the individual and the population. Spread of disease is of importance for both, as medical epidemiology and social epidemiology possess medical and behavioral elements that provide a balance and can result in greater insights when combined together.

Living on Mars will present unique challenges, as the physical environment is inhospitable to human biology and that makes the social environment, the ecosystem inside, challenging as well. The unique Martian living conditions will present physicians and medical astrosociologists with unique issues with which they must seek to understand and then produce life-saving and/or sociocultural and psychological responses. The concept of behavioral health provides a useful approach that involves both branches of science.

13.5 Mitigating deleterious effects on Mars

Several proposals have arisen as possible solutions to mitigate the deleterious effects produced by living on Mars. For example, terraforming the Martian environment is one option to reduce health risks. However, it is tremendously difficult, time consuming, requires tremendous resources, and thus not everyone agrees that it is the best strategy to follow. **Figure 6** depicts what three stages of a terraformed Mars may look like. This is not necessarily a likely solution to living on Mars for the foreseeable future and must overcome ethical objections to even starting such a planetwide procedure. The process would wipe out any existing Martian life, even fossils that may exist, and it could add new detrimental health problems for humans. There are arguments on both sides of this issue.

Terraforming Mars would involve releasing greenhouse gases, what on Earth are regarded as pollutants, into the Martian atmosphere from a number of different

Figure 6.
Terraforming Mars/image credit: NASA.

locations on the planet in order to thicken the atmosphere and make it breathable. Such a herculean effort would take hundreds if not thousands of years. The resource expenditures and length of the process have led proponents of Martian settlement to seek other options.

13.6 Genetics and lava tubes

Genetic enhancements may better adapt humans to living on Mars physically speaking, but what will it do to the human psyche? How will it affect behavior? Is it even ethical? What about the history of eugenics? The Nazi regime in Germany instituted various forms of eugenics, which led to unethical experiments and genocide against Jews and other perceived unworthy categories of people. While altering the human body to better tolerate solar radiation, for example, may seem like a logical idea, the human experience in this area is highly problematic. Ethical questions abound such as to whether or not those who receive enhancements have clearly provided authorization.

Other proposals include living inside lava tubes, as mentioned, utilizing 3D printing to construct habitat components, and creating human made subterranean habitats. While the external environment that impacts the surface is harmful to human life that currently exist, the places where humans actually live—that is, the ecosystems in the habitat structures—present potentially equally dangerous conditions as well.

13.7 Dangers in the habitat

Life within the habitat structures is also full of health and safety threats. Simple survival requirements also present significant challenges that include providing water for humans and crops, creating an ongoing supply of oxygen, growing food in Martian soil and/or hydroponically, interacting with others on a prolonged basis within an confined and enclosed space, keeping Martian dust out of the habitat, dealing with circadian rhythm issues, adjusting to time differences (the Martian day is 40 minutes longer than Earth's), and dealing with delayed communications with Earth. Accumulating challenges could produce problematic behavioral and abnormal psychological patterns for some individuals.

Bringing Martian microbes into the laboratories within the habitat can present contamination worries. It can result in microbes escaping quarantines or otherwise infecting astrobiologists. An epidemic could arise if the microbes are harmful to humans.

Physical epidemiologists and social epidemiologists would have to work together to track the spread of disease and the structure and procedures employed to combat its spread. There are other issues that can result in health-related problems generated in the outside environment that have both medical and sociocultural implications.

13.8 Aging and space gerontology

Another issue is less theoretical and actually inevitable. Space medicine will be difficult enough without the added complications of advanced aging. Many settlers will necessarily be older based on the need for them to put their expertise and training into effect. In any case, they will age. Therefore, gerontology, defined as the scientific study of old age, the process of aging, and the particular problems of old people, will become important as everyone ages under a never-before set of medical conditions exacerbated by the Martian environment. It will involve medical astrosociology in terms of how individuals handle growing old on Mars in addition to the biomedical changes that will challenge them.

Space gerontology has been recognized by NASA and others for quite some time, in fact since 1978 or earlier. The advent of the Space Shuttle program with its first flight of Columbia in 1981 and the recognition that older astronauts would be traveling into near Earth orbit in microgravity, it became worrisome that little was known about the long-term effects of space travel as people age [69]. Nearing the end of the human life cycle in space ecosystems will produce changes, many of which will have never been experienced on Earth. Will ageism be a problem if, for example, not enough medical supplies exist for the younger members of society.

How will settlers handle death? What will cemeteries on Mars look like and what types of religious ceremonies will be held? What meanings will be attached to the deaths of the new Martians? It is quite possible that the ways that deaths are treated will evolve into one or more uniquely Martian versions of Earth observances. This type of sociocultural change is important for astrosociologists to study and understand.

Very importantly, the added complications affecting human beings in a harsh Martian environment that will inevitably impact significantly on the internal ecosystem will make implementing healthcare extraordinarily unique and difficult. What was learned regarding living in microgravity will not apply exactly the same way in a one-third gravity field. Gerontologists who also possess social science training will prove to be invaluable members of the Martian society, just as will geriatric physicians.

13.9 A few final questions and thoughts

What new syndromes or health complications will arise on Mars? How will living on Mars affect men compared to women? Will health issues affect ethnic groups differently? Will life expectancy change? These types of questions must be anticipated by both physicians and medical astrosociologists before arriving at Mars; and more importantly, monitored once the settlement is active.

While space medicine physicians focus mostly on the biomedical issues, medical astrosociologists concentrate on the social and cultural elements that interact with the biomedical aspects that can easily add further complications beyond the strictly medical concerns. This complicates how those reacting to the patient should handle any particular case. For example, are there religious factors that do not allow certain medical procedures or drugs? How do physicians and social scientists work together to treat patients, avoid discriminatory practices, and monitor the population to minimize epidemic potentials in such enclosed ecosystems? Medical astrosociology on Mars is required in concert with space medicine and behavioral health in order to provide successful Martian healthcare. Medical and social-scientific complications will abound in such scenarios. The relationships among those practicing space medicine and medical astrosociologists will also prove to be challenging.

14. From space law to Martian law

Laws are legal norms or officially specified rules of behavior. "Outer space law encompasses both international and national law related to all aspects of space technologies, human and robotic activities and conduct, and the applicability of such laws to outer space environments and ecologies as specified in treaty, statute, or code [70], p. 3." The Outer Space Treaty (OST) and Moon Treaty were drafted to set the rules of behavior for activities beyond the Earth. They have substantial problems that seem to increase in seriousness and scope as space activities continue to increase. Relatedly, some nations such as the United States have drafted legislation that counters the intent of the treaties. If treaty provisions are not honored among terrestrial nations, what hope is there for Martian settlements to honor them?

Just as on Earth, law enforcement and social program efforts will need to react—or more importantly, if possible, prevent—violations of norms, both legal and administrative, that include being drunk in public, burglary, health and safety offenses, homicides, and a host of other behaviors that commonly occur on Earth. Humans will not act altogether differently just because they transfer to a Martian settlement. In fact, the isolation and confinement during the voyage may result in an increase in deviant behavior. Planners must therefore prepare for well-known outcomes that human societies have experienced throughout history. This is a starting point.

Moreover, what happens when settlers begin to modify or even reject treaties and laws drafted on Earth that they deem impinge on the evolving values and norms in Martian settlements? Earth-based space law exists for the benefit of Earthlings, but not all of them will apply for Martian settlers who will find themselves facing very different social forces and realities. Once humans begin to construct the physical habitat components for a particular settlement and individuals start to inhabit them, the laws established before the trip to Mars begin to evolve. (This will likely occur during the voyage to Mars).

Because laws are tied to the societies in which people actually live, they are drafted and enforced according to societal realities unique to them. On Earth, various nations enact laws that differ in many ways from those of others based on cultural identity and political priorities. Martian settlements will face uniquely challenging social forces and conditions. Thus, sociocultural change and potential social movements will produce modifications in the law. For example, will the property rights prohibition in the Outer Space Treaty be ignored or legislated as nonapplicable by settlement governments? Predictably, settlement laws will quickly start to diverge from the Earth-based case law, treaty provisions, and the traditions of terrestrial societies.

15. Planetary protection

"Astrosociology deals with the broad, societal contexts of activity pertaining to space, as well as actual space exploration including human space exploration and the search for extraterrestrial life [71]." The reference to extraterrestrial life is important because the search for life beyond Earth—especially on planets such as

Mars and Venus, and moons such as Europa, Ganymede, Enceladus, and Titan—largely depends on taking measures to avoid contaminating potential extraterrestrial ecosystems, extant life, as well as fossils. In fact, the Outer Space Treaty has a provision that emphases protection of outer space bodies [72].

"Planetary Protection is the practice of protecting solar system bodies from contamination by Earth life and protecting Earth from possible life forms that may be returned from other solar system bodies," according to the NASA Office of Safety and Mission Assurance [73]. Sterilization of spacecraft and habitat components is vital for Martian settlers with astrobiological backgrounds to make sure that they do not introduce Earth organics that may contaminate potential Martian organisms. Mistaking Earth microorganisms for Martian ones is something that they want to avoid. Others may not worry nearly as much about this issue, so astrobiologists would favor legal prohibitions that protect the potential Martian life so that proper precautions are taken by everyone. It must be taken seriously if astrobiological science is to be protected from harmful events and needless mistakes by technicians.

Placing a settlement on Mars also involves potential limitations regarding where to locate. For example, a water source would be ideal for obtaining water, but this is also where microorganisms are likely to exist. Conflicting priorities may well exist between astrobiologists and government officials. In a best-case scenario, water drilling operations that provide water for the settlement would occur in concert with attempting to limit the amount of environmental damage so that the scientific objectives of astrobiologists and others who want to protect potential Martian life can occur. Landing or placing buildings in areas thought to be good candidates for life would violate the concept of planetary protection. It will not be possible to protect every part of Mars thought to be a candidate for life. Moreover, economic commerce between Earth and Mars will place added pressures to protect the ecosystems of both worlds.

16. Exo-astrosociology on Mars

Astrobiology and exobiology represent the physical and natural science approaches to the study and search for extraterrestrial life. While the distinctions and practices that separate these two fields are somewhat murky, one general expression that does so is as follows. Differences exist historically about distinguishing between astrobiology and exobiology, so a redefinition is provided here. Astrobiology seeks to find a second genesis of life within our own solar system while exobiology seeks to find it beyond our solar system, the latter of which has become more relevant with the increasing discovery of exoplanets and more recently exomoons. These are by far the most common approaches to the search for extraterrestrial life and they are specialized in the natural and physical sciences rather than the social sciences. Practically, astrobiology tends to be the overarching title for the search for all life.

Nevertheless, as with other issues, there are astrosociological implications involved with the search for extraterrestrial life. The social sciences have taken a back seat, historically, although things are changing and there were always exceptions to this general rule. For example, a NASA publication entitled "Workshop on the Societal Implications of Astrobiology" was published in 2001, which clearly focused on the importance of the social sciences and included social scientists [74]. This provided a bedrock for future off and on collaboration.

However, an academic field dedicated to the focus on the social-scientific implications of the search for extraterrestrial life was needed. "A new discipline dubbed 'astrosociology' has arisen in the past few years that addresses the societal impact of space exploration, including extraterrestrial life [75], p. 174." Thus, there are social implications of astrobiology [76]. "The social and cultural implications of this [astrosociological] work make it too important to ignore. In fact, it is imperative that astrosociologists participate alongside their space-community counterparts to attain comprehensive knowledge; both for its own sake and for practical application should some type of reaction prove necessary [75], p. 175." This is the case on Mars or wherever a discovery is made.

Exo-astrosociology, which was first introduced by this author, is a recently created astrosociology subfield. It is defined as the study of extraterrestrially-related forms of astrosocial phenomena. It "involves how the very search for extraterrestrial life impacts humanity in a myriad of different ways" and "also involves how ongoing failure and potential success affects societies, cultures, social groups, subcultures, and individuals [77], p. 4." This subfield was finally introduced in order to create a community of like-minded social scientists and a way for astrobiologists to more easily collaborate with them. An exo = astrosociology literature will add a missing component in a more organized fashion than existed before its inception.

Mars is good place for conducting SETI and astrobiological research. Of course, life may well exist on the surface or subsurface of Mars and biosignatures likely exist. The concept of planetary protection, if carried out as conceptualized and described above also makes it possible to conduct research to discover extant and fossilized life. The search for life beyond Mars is also an improvement compared to Earth in some ways. The atmospheric volume is only one-percent of that of Earth's, which provides a good environment for surface-based telescopes when dust storms are not obscuring the view. It is also possible to place telescopes on the moons of Mars, Deimos and/or Phobos. There are several possibilities that require investigation.

Looking for a second genesis on Mars will be a priority for scientists. Where water exists, there could also exist Martian life. The existence of extremophiles on Earth consist of organisms that live in extreme environments such as "(1) in Mono Lake (a highly alkaline environment) (see **Figure** 7), (2) in Yellowstone's Sylvan Springs (sulfuric acid), (3) within caves deep under-ground in harsh conditions, (4) within Antarctic ice sheets (including Lake Vostok), and (5) thriving on highly radioactive control rods in nuclear power plants" [78, pp. 402-403]. Extremophiles on Earth provide optimism that similar extant or fossilized microbes can be discovered on Mars. In fact, the choice to live in lava tubes may bring about objections by astrobiologists and exo-astrosociologists due to the possibility of contaminating life and destroying fossils in these caves. Those searching for life are excited to explore potential nonhuman Martian ecosystems above the surface and below while protecting it from contamination and destruction to the extent possible.

As discussed, water is an important resource to sustain human life as well. Mining will be a priority for others. Water exists at the poles, but it also exists elsewhere at much lower latitudes (see **Figure 8** as an example) [79]. Drilling for water will occur. Putting in place the infrastructures to locate water and transfer it to habitat components is a messy enterprise. Ideally, settlements will be placed relatively close to sources of water. Thus, when humans finally live on Mars, planetary protection will become increasingly difficult, as we have witnessed on Earth when human communities encroach on previously pristine ecosystems. How will competing factions work out this clash of interests? It is important for astrosociologists to be there to study these types of dynamics and perhaps assist in resolving tensions.

17. Space archeology on Mars

Space archeology is defined as the study of human-made items, or material culture, found in space environments (including on Earth) and their relationships

Figure 7.
Mono Lake.

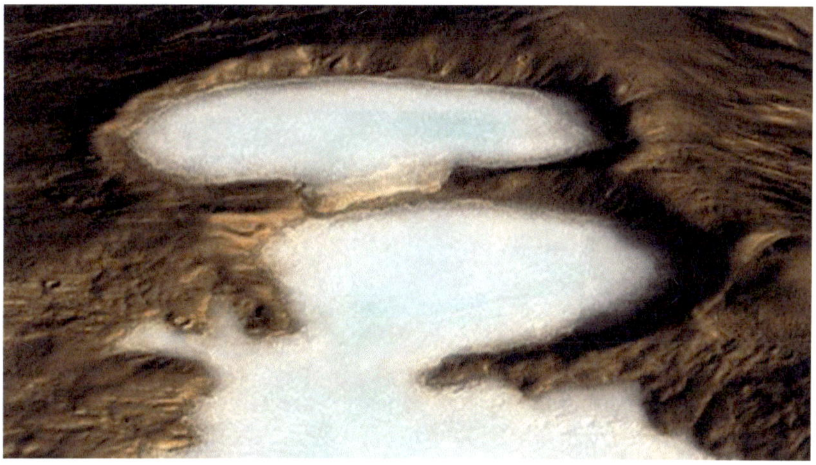

Figure 8.
Artist concept of a glacier on Mars discovered in 2008. Photo credit: NASA/JPL.

to human exploration, which includes efforts to preserve them as cultural heritage [80]. Interestingly, space archeology has only gained acceptance in the last decade or so according to Dr. Alice Gorman [81]. Dr. Gorman and other space archeologists focus on protecting artifacts on the Moon in orbit around the Earth, and material culture on Earth's surface including space debris. Examples include the Apollo 11 site and others visited by humans, rovers, and landers. The same arguments apply to protecting the heritage of sites on Mars as well, and wherever humans leave their material culture. Preserving history for generations represents a laudable goal.

Beyond protection of potential life is that of the material culture, potential monuments, that rest on the Martian surface. These emissaries of humankind include the Viking landers, the Pathfinder lander and Sojourner rover, the Spirit and Opportunity rovers, the Phoenix Science Laboratory, the Curiosity rover (see **Figure 9**), and the InSight lander. Interestingly, the Spirit Rover ended operations on the north face of the Troy plateau in January of 2009. Crash sites could also be of value, as the debris

Figure 9.
Curiosity snaps a belly selfie at Buckskin Mountain base drill site. Photo credit: NASA/JPL.

may finally help determine causes of the failures. One example is the Mars Polar Lander. The Beagle 2 landed successfully and began some preliminary operations, but it shut down and failed to contact Earth. Protection of these historically significant items may seem like an obvious way to handle them, but history has shown that some individuals opt to grab pieces for souvenirs and others simply favor destroying them. Scientists may also abscond with pieces of rovers and other artifacts for analyses of various types. Destruction of artifacts, or at least serious damage to them, is possible or even probable without adequate protections and Martian laws in place.

Legal prohibitions may not allow direct access to heritage sites with some logical exceptions, but adherence to such laws may be unheeded by some settlers. As with other areas of social life, social and behavioral scientists need to study social, cultural, and behavioral patterns that can inform policy makers and law enforcement officials of violations. They can also help to design tourist venues assuming that they are practically accessible by settlers while also being protected. Study of this behavior would also be of value to exo-astrosociologists and therefore the Martian settlement.

18. Interplanetary relations

It cannot be emphasized enough that settlers on Mars (or anywhere off the planet Earth) will adapt to their physical environment within the habitat. They will seek to protect themselves from health and safety risks that impinge on them from the larger environment beyond its walls such as radiation, an altered gravitational field, and the inability to breathe without a spacesuit on the Martian surface. This is because isolation and confinement alone will impact on both physiology and behavior [67]. These factors and several others will cause human populations to adapt to these physical factors, which will result in social forces that will inevitably modify the base culture set up during the planning stages on Earth and that have been modified during the voyage. For the settlement, social and cultural change will continue from the training on Earth to the landing on the Martian surface, and beyond that point [44].

A significant part of this process will alter the culture in ways that eventually cause residents to view themselves as members of the Martian settlement apart from their identities as humans from terrestrial societies such as the United States or Russia. A socialization process will develop in which a group mentality reinforces their Martian or specific settlement identity over time. For those who stay permanently, this process will occur due to individuals interacting with others around them rather than relying on past memories of their time before settling on Mars.

Dependence on Earth-based governments and private organizations for supplies can take its toll as can difficulties during their interactions. Group cohesion among settlers will develop and increase over time, and ethnocentrism could well result in them thinking of themselves as "different." In such a situation, they will become part of their in-group and Earth dwellers will become the out-group, at least to some extent. At first, settlers will require assistance in many forms including food, water, and other supplies in order to survive. However, to the extent that they can fend for themselves over time and provide goods and services unique to Mars, they may be able to wean themselves off of their dependency from relying on Earth for at least *all* of their requirements.

These processes may take quite a bit of time although we have seen these patterns on Earth in the development of nations, including their realignments and border changes. Changing cultural norms and values can result in dramatic changes of fundamentally held ideas and the resulting changes in behavior. Ideas can forge new social institutions and behavioral patterns of the larger population that can become very different than when the social system started.

Economically, humans have been used to transacting with other Earth-based partners. However, for settlers on Mars, Earth-to-Earth transactions for which they were familiar will have changed to an Earth-to-Mars relationship. In this scenario, the supply run from an Earth-based nation, corporation, or other entity to an extraterrestrial human settlement would change for both those on Earth and those on Mars. For settlers, to the extent that leaders of the Martian settlement perceive they are being mistreated in some ways, this can escalate the path toward separation of the settlement from all forms of Earth-based jurisdictions. They would, in this case, behave in accordance with their own laws.

The process of getting to a point in which this type of relationship between the two planets manifests is theoretically a long one. However, it can be accelerated to the extent that leaders evaluate the situation in a way that they feel being exploited. This becomes a shift to *interplanetary relations* – which is akin to the international relations found on Earth – that takes place between (1) Earth-based governments, non-governmental organizations (NGOs), corporations, and other entities, and (2) a particular Martian settlement.

19. Martian planetary defense

Planetary defense efforts have logically focused on protecting against possible strikes against Earth. Moreover, it has increased in seriousness over the last couple of decades. The B612 Foundation was founded in 2002 to provide a nongovernmental voice focusing on the need to protect Earth from asteroids and comets [82]. Other space advocacy organization such as The Planetary Society [83] and The National Space Society [84] also provide resources and efforts related to improving planetary defense resources and capabilities. NASA has been searching for near Earth Objects (NEOs) since 1998 when Congress first directed it "to find and track 90 percent of NEOs 1 kilometer or more in diameter within 10 years, as objects of this size would cause a global catastrophe [85]." The newly created Space Force is slated to take over the responsibility of finding and cataloging NEOs [86]. In addition to NASA, the European Space Agency (ESA) is also taking part in locating NEOs [87]. The search for threatening asteroids and comets is serious business to which both public and private entities contribute. Recent activity to protect Earth from a catastrophic event demonstrates how important this issue has become.

If it is serious enough for terrestrial nations to take action, the prospects for other locations in our solar system are even more dire. Both cosmic bodies first to be settled, the Moon and Mars, are more susceptible to asteroid and comet fragment strikes than the Earth due to its much denser atmosphere. While the lunar atmosphere is negligible, the Martian atmosphere does exist though projectiles from space do not burn up as readily as is the case for Earth. Because the Martian atmosphere is so thin as discussed earlier, we know from spacecraft sent there that objects hit hard. "Even though the surface gravity on Mars is only 3.7 meters/sec (compared to 9.8 meters/sec on Earth), the thin atmosphere means that the average terminal velocity hits a nail-biting 1000 km/hour or so, compared to about 200 km/hour back home [86]." While landing on Mars is difficult due to its thin atmosphere, as exemplified by spacecraft having to do a lot of work to shed speed, asteroids and comets do not need to "worry" about that although settlers sure need to worry about being hit at such great speeds (see **Figure 10**).

Thus, based on activities on Earth, the need to keep track of Near Mars Objects (NMOs) is perhaps of greater concern. Mars is a smaller planet, but its atmosphere is much less dense, so a direct strike would be more devastating for an object of the same size. It is important to put things in perspective, however. The chances of a cosmic object striking a settlement is very low. It would increase with a larger perimeter of settlement components, but it would still be very unlikely to occur. Nevertheless, the Martian surface is full of craters caused by strikes from space. Therefore, keeping watch wherever humans reside is a good idea. Earthlings do it from their home planet and aboard the International Space Station (ISS). Infrastructure to deal with possible comet or asteroid strikes would need to be constructed as part of the settlement. Telescopes orbiting Mars, perhaps on one or both of its moons, or in nearby space, would mimic what humans do, or will do, on Earth.

As on Earth, this authored has proposed that there are three main phases related to coping with the possibility of an asteroid or comet strike: (1) detection, (2) defense, and (3) survival [88] (see **Table 1** for details). The detection and defense components of the planetary defense strategy are quite obvious and they receive by far the most coverage by scientists the information media, social media, and entertainment outlets. A potential strike of Earth is much more

Figure 10.
Artist illustration of an object striking Mars/photo credit: Don Dixon.

(1) Detection	This component of the strategy involves the detection of possible Mars-bound asteroids and comets in orbits and trajectories that threaten to collide with Mars.
(2) Defense	Once a threat is verified to be on a collision course with Mars and a potential threat to a Martian settlement, either directly or indirectly, this component involves eliminating the threat using whatever means are available.
(3) Survival	Should the second component fail on a partial or complete basis, this component involves moving people and vital resources (if possible) to safer places such as lava tubes.

Table 1.
Summaries of the three components of the planetary defense strategy (modified from Pass [88], p. 8).

serious than a Martian threat due to the population differences although either scenario could be catastrophic.

Even though the threat is less likely to threaten a Martian settlement, the possibility still exists. What happens if the defense efforts fail? In such a case, if the population is endangered, the survival of the settlers becomes the priority. Moving people to a safer area is easier because the population is relatively small compared to Earth and the unpopulated areas are vast. The problem that does exist is that individuals would need to be suited up in order to survive outside. Rovers could provide much needed methods of escape. As on Earth, preplanning to better cope with such a disastrous scenario is the optimal course of action.

20. Conclusions

Too many important astrosociological issues exist to cover in a single chapter, although many are found here, which has resulted in cursory coverage of some areas. It does demonstrate, however, how many issues require astrosociological research so much more astrosociological research is needed. This means that many more trained astrosociologists must exist to accomplish it. This will require a substantial increase in astrosociology education that will, in turn, result in an increase in related research. While an astrosociology course is being taught at Harvard University, it is imperative for astrosociological education to spread throughout academia in order to settle Mars successfully. ARI's "Astrosociology in the Classroom" program is important for a number of other reasons as well. This discussion about a hypothetical Mars settlement demonstrates the fact that too little knowledge currently exists.

Thus, while it is true that no human has yet landed on the Martian surface, it is not too early to think about it and study these types of astrosociological issues because the underpinnings of planning for a Martian settlement already exist and they are becoming increasingly prevalent. The movement of lunar and Martian human exploration and permanent settlement has spread to the private sector with the advent of NewSpace companies and nonprofit organizations. Space advocacy groups proliferate. Progress in support of space exploration continues. In order to further this movement, more space professionals and students versed in both the physical and social sciences need to access their astrosociological imagination. They need to understand the true impact of astrosocial phenomena in their lives and in their societies. Only then will they better recognize the value of working together rather than separately as they have too often in the past.

Many complex issues are discussed in this chapter. Planning for all three stages (i.e., on Earth before launch, during the transit phase, and once arriving at Mars) represents a crucial approach that must involve all types of scientists and scholars who are relevant contributors to space exploration and living in isolated and confined spaces. There are both physical and social environments that involve architecture and art, cultural considerations, and the construction of social institutions beyond construction of spacecraft and habitat components. Deviance and other social problems will manifest on a continuing basis that will require astrosociological research and intervention. Other areas relevant to settling on Mars that are covered here include medical astrosociology, space law to Martian law, planetary protection, exo-astrosociology, space archeology, interplanetary relations, and Martian planetary defense.

All of these space-related social-scientific issues – astrosocial phenomena – reflect a type of inquiry that requires significantly greater attention by social scientists. The hope here is that these issues are brought to the forefront so that future Martian settlements are successfully constructed and that the social systems in which their citizens live can thrive beyond simple survival, whether they become realities in 10 years or further down the line. Otherwise, rejecting astrosociology on Mars will result in hardships and social problems that could have been avoided or at least greatly mitigated by astrosociologists.

The Astrosociology Research Institute exists to make astrosociology on Mars possible. As emphasized throughout this chapter, the settling of Mars involves many issues that transcend those covered by STEM disciplines and fields alone. Astrosocial phenomena include realities that focus on the human dimension, which goes beyond the concerns of physical and natural scientists. However, these individuals often make remarks about social, cultural, behavioral, and psychological issues. At the same time, most of them have not studied these types of issues in any great detail although it demonstrates the connection between the physical and social sciences. These types of scientists as well as humanities scholars and artists are needed because they are trained in these subjects. Together, when they collaborate, they can provide expertise from both branches of science to elicit insights that neither side alone can produce. As **Figure 11** illustrates, astrosociology is a multidisciplinary, collaborative scientific field, and convergence among all relevant disciplines and fields represents a major objective that is currently on the path to significant realization.

It is unwise for planners of a Martian settlement to move forward without social-scientific input, if individuals visit the settlement as tourists and especially if they commit to staying permanently. There are too many statuses and roles that need to depend on one another, which means that planners must focus on the intricacies

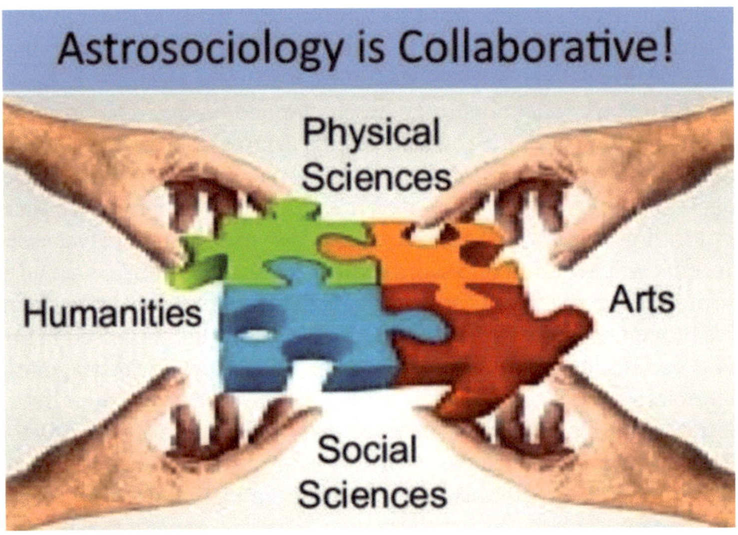

Figure 11.
Astrosociology is a multidisciplinary academic field.

involving constructing a functional social system—especially as its population grows—just as the greatest attention is traditionally paid to the construction of physical assets such as rockets and the habitats and their systems. Like the engineering systems in a spacecraft, for example, a Martian city or smaller settlement, possesses its own intricate systems in the form of human relationships. The latter is more difficult to plan for and keep functioning in a positive way because the components, in this case humans, keep changing unlike what occurs to the same extent in physical systems.

So, while rocket science can get people to Mars and engineering can keep the habitat functioning, it cannot ensure societal success within the habitat. The crux of the matter boils down to one question. What will happen to human life if astrosociology fails to get to Mars? Without a semblance of social order and cultural consistency, rocket scientists or mathematicians can do very little to construct a stable Martian society. Both branches of science must work together. Only cooperation and collaboration between those in both branches can result in a successful outcome on a sustainable basis. Physical and social systems must work together, as they are complementary, and that requires physical and social scientists to work together.

This chapter has focused on some of the major issues relating to Martian settlements, and perhaps some of them are not frequently discussed as part of this larger theme, but it will be the first location to be settled beyond the cislunar environment and it is important to expand our astrosociological imagination beyond what is common. Moreover, the difficulties of humans seeking to settle in other space environments will only become more pronounced the farther away from Earth that they migrate. Social life will become increasingly difficult in a number of social, cultural, and psychological dimensions. If human settlements can successfully transform themselves into well-functioning space societies, they will provide those who migrate beyond Mars with invaluable lessons of survival. Additionally, the social problems and failures that become documented will also provide these intrepid settlers with lessons of what to avoid if possible. Thus, it is important for astrosociologists to be among the settlers in order to monitor, document, and support social life in Martian settlements and beyond.

It is not logical to exclude those trained in the social sciences. They are important on Earth, so why would they be less so on a planet inhospitable to human life, which challenge settlers socially, culturally, and psychologically? They are important wherever humans reside, whether here on Earth or beyond. Astrosociology on Mars makes logical sense because astrosociology on Earth is also invaluable to human societies and their citizens because they offer complementary insights to those in the physical and natural sciences who traditionally study space settlement issues.

Thanks

I would like to thank two late individuals who inspired me, Dr. Allen M. Tough (1936–2012) and Dr. Albert A. Harrison (1940–2015), as a reminder to all those who support astrosociology. Allen Tough was a futurist and SETI scholar who mentioned the term "astrosociology" as a possible new social science field in an article that prompted me to found the field immediately after reading it [26]. I met him one time in person at a conference and he encouraged me to continue my efforts. Albert A. Harrison was a social psychologist who I met online initially after reading one of his books [27]. He became a friend and the first advisor to The Astrosociology Research Institute. He attended several conferences with me and we cowrote a few publications. His early inspiration in 2005 helped me greatly to carry on when things were not always easy. These two individuals inspired me the most when I decided to found astrosociology in 2004.

I would also like to thank Dr. Thomas Gangale and the late Dr. Marilyn Dudley-Rowley who I met in 2004 at a sociology conference. They were early supporters and participants in astrosociological efforts including helping me to found The Astrosociology Research Institute. I also wish to thank Dr. Gangale for creating ARI's logo (see **Figure 2**).

Author details

Jim Pass[†]
The Astrosociology Research Institute, Huntington Beach, CA, USA

*Address all correspondence to: jpass@astrosociology.org

† This author is also available on Twitter (twitter.com/astrosociology), Facebook (facebook.com/jimpass2), and LinkedIn (linkedin.com/in/jimpass/).

References

[1] Pass J. Astrosociology: Exploring the human dimension of outer space. Ad Astra. 2019;**3**:43-44

[2] Why Social Science? Because We Will Need to Do Better in the Next Crisis. Available from: https://www.whysocialscience.com/

[3] Pass J, Harrison AA. Chapter 38: Astrosociology (social science of space exploration). In: Bainbridge WS, Roco MC, editors. Handbook of Science and Technology Convergence. New York: Springer; 2016. pp. 545-558

[4] Pass J. Astrosociology, education and the future of space exploration, exploitation, and settlement. In: Paper Presented at the AIAA Space Aeronautics Forum and Exposition. Paper Written for the Society and Space Session for the AIAA Space Conference and Exposition; 12-14 September 2017; Orlando, FL. 2017. p. 2. Available from: http://www.astrosociology.org/Library/PDF/Space2017-Pass-AstrosociologyEdAndFuture.pdf

[5] Pass J. Target viewpoint: Examining the definition of astrosociology. Astropolitics. 2011;**9**(1):6-27

[6] Peters MA. Beyond Earth's globalization: Sociology of outer space? Review of Contemporary Philosophy. 2017;**16**:83-91

[7] Pass J. Inaugural Essay: The Definition and Relevance of Astrosociology in the Twenty-First Century (Part One: Definition, Theory and Scope). 2004. Available from: http://www.astrosociology.org/Library/Iessay/iessay_p1.pdf

[8] Pass J. Space: Sociology's forsaken frontier. In: Presented at the California Sociological Association (CSA) Conference; 16 October 2004; Riverside, CA. Available from: http://www.astrosociology.org/Library/PDF/submissions/Space_Sociology%27s%20Forsaken%20Frontier.pdf

[9] Finney BR. From Sea to Space (The Macmillan Brown Lectures, 1989). Honolulu: University of Hawaii Press; 1992

[10] McKay CP. What is life – And when do we search for it on other worlds? Astrobiology. 2020;**20**:163-166

[11] Boston P. The Case for Mars: Proceeding of a Conference Held 29 April-2 May 1981 at the University of Colorado, Boulder (Science and Technology Series). American Astronautical Society; 1984

[12] Shostak S. Introduction: The true nature of aliens. International Journal of Astrobiology. 2018;**17**(4):281

[13] Harrison AA, Clearwater YA, McKay CP, editors (Forward by Gunderson EKE). From Antarctica to Outer Space: Life in Isolation and Confinement. New York: John Wiley & Sons, Ltd.; 1989

[14] Finney BR, Jones EM. Interstellar Migration and the Human Experience. Berkeley: University of California Press; 1985

[15] Swenson LS Jr, Grimwood JM, Alexander CC. This New Ocean: A History of Project Mercury. Washington, DC: National Aeronautics and Space Administration (NASA); 1998

[16] Kanas N. Humans in Space: The Psychological Hurdles. Springer Praxis Books; 2015

[17] Vakoch DA, editor. Psychology of Space Exploration: Contemporary Research in Historical Perspective. Washington, DC: National Aeronautics and Space Administration (NASA); 2019

[18] McCurdy HE. Space and the American Imagination. Baltimore: John Hopkins University Press; 1997

[19] Launius RD. Frontiers of Space Exploration. 2nd ed. Westport, CT: Greenwood; 2004

[20] Logsdon JM, John F. Kennedy and the Race to the Moon (Palgrave Studies in the History of Science and Technology). London: Palgrave Macmillan; 2011

[21] McCurdy HE. Financing the New Space Industry: Breaking Free of Gravity and Government Support (Palgrave Studies in the History of Science and Technology). New York: Palgrave Pivot; 2019

[22] Finney B. Lunar base: Learning to live in space. In: Mendell WW, editor. Lunar Bases and Space Activities of the 21st Century. Washington, DC: National Academy of Sciences; 1984. pp. 751-756

[23] Bluth BJ. Lunar settlements: A socio-economic outlook. Acta Astronautica. 1988;**17**(7):659-667

[24] Harris PR. Behavioral science space contributions. Behavioral Science. 1989;**34**(3):207-227

[25] Rudoff A. Societies in Space (American University Studies, Series XI, Anthropology & Sociology. Vol. 69). New York: Peter Lang Publishing; 1996. p. 75

[26] Tough A. Positive consequences of SETI before detection. *Acta Astronautica*. 1998;**42**(10-12):745-748

[27] Harrison AA. Spacefaring: The Human Dimension. Berkeley, CA: University of California Press; 2001

[28] Dudley-Rowley M. The great divide: Sociology and aerospace. In: Paper Presented as Part of the Special Dedicated Session Entitled "Astrosociology: The Establishment of a New Subfield" at the California Sociological Association (CSA) Conference; 16 October 2004; Riverside, CA. 2004. Available from: http://www.astrosociology.org/Library/PDF/submissions/The%20Great%20Divide_CSA2004.pdf

[29] Sadeh E. Chapter 8: Impacts of the Apollo program on NASA, the space community, and society. In: Dick SJ, editor. Historical Studies in the Societal Impact of Spaceflight. Washington, DC: National Aeronautics and Space Administration (NASA) History Program Office; 2006. pp. 491-534. Available from: http://www.astrosociology.org/Library/PDF/Historical-Studies-Societal-Impact-Spaceflight.pdf

[30] Gangale T. Common heritage in magnificent desolation. In: Paper Presented as Part of a Session Entitled "Astrosociological Perspective on Space Exploration" at the AIAA Aerospace Sciences Meeting and Exhibit (ASM 2008) Conference; 8 January 2008; Reno, NV. 2008. Available from: http://www.astrosociology.org/Library/PDF/ASM2008_MagnificentDesolation.pdf

[31] Lockard ES. Symbiocracy: The structuring of new societies in space based on principles of mutualism and symbiotization. In: Paper Presented as Part of a Session Entitled "Astrosociological Perspectives on Space Societies" at the AIAA Space 2008 Conference; 10 September 2008; San Diego, CA. 2008. Available from: http://www.astrosociology.org/Library/PDF/Lockard_Symbiocracy.pdf

[32] Bainbridge WS. Virtual world astrosociology. In: Proceedings of the 2nd Symposium on Astrosociology that was part of the Space Propulsion and Energy Sciences International Forum (SPESIF) Held at the Johns Hopkins University Applied Physics Laboratory, Laurel, Maryland. The session was called "Overview Effect". American Institute of Physics; 2010. Available

from: http://www.astrosociology.org/Library/PDF/SPESIF2010_Bainbridge_VirtualWorld.pdf

[33] Hearsey CM. The nexus between law and astrosociology. Astropolitics. 2011;**9**(1):1-5

[34] Lempert D. Living in space: Cultural and social dynamics, opportunities, and challenges in permanent space habitats. Astropolitics. 2011;**9**(1):84-111

[35] Pop V. Space exploration and folk beliefs on climate change. Astropolitics. 2011;**9**(1):50-62

[36] Dator JA. Social Foundations of Human Space Exploration (SpringerBriefs in Space Development). Cham, Switzerland: Springer International Publishing; 2012

[37] Oman-Reagan M. Space exploration and planetary perspectives. In: Schultz EA, Lavenda RH, Dods RR, editors. Cultural Anthropology: A Perspective on the Human Condition. 4th Canadian ed. Canada: Oxford University Press; 2018. pp. 296-297

[38] Wood J, Schmitt L, Lugg D, Ayton J, Phillips T, Shepanek M. Life, survival, and behavioral health in small closed communities: 10 years of studying isolated Antarctic groups. Aviation, Space, and Environmental Medicine. 2005;**76**(Supplement 1):B89-B93

[39] Kanas N. Chapter 3: Psychosocial value of space simulation for extended spaceflight. In: Bonting SL, editor. Advances in Biology and Medicine. Vol. 6. 1997. pp. 81-91

[40] Palinkas LA. Psychosocial issues in long-term space flight: Overview. In: Gravitational and Space Biology Bulletin. 2001;**14**(2):25-33

[41] Tachibana K. A Hobbesian qualm with space settlement. Futures. 2019;**110**:28-30

[42] Mills CW. The Sociological Imagination. New York: Oxford University Press; 1959. p. 6

[43] Pass J. Albert A. Harrison: Outer space, the human dimension, and astrosociology. In: Paper Presented in the Society and Space Session at the AIAA Space Conference; 13-16 September 2016; Long Beach, CA. 2016. p. 4. Available from: http://www.astrosociology.org/Library/PDF/Space2016-JPass-AlbertAHarrison.pdf

[44] Pass J. Astrosociology and the planning of space ecosystems. In: Paper Presented as Part of the Space and Society Session at the Space 2015 Conference; September 2015; Pasadena, CA. 2015. Available from: http://www.astrosociology.org/Library/PDF/Space2015--JPass-PlanningSpaceEcosystems.pdf

[45] Hancock PA. On bored to Mars. The Journal of Astrosociology. 2017;**2**:103-120

[46] Blackwell-Landon L, Slack KJ, Barrett JD. Teamwork and collaboration in long-duration space missions: Going to extremes. American Psychologist. 2018;**73**(4):563-575

[47] Pass J. The astrosociology of space colonies: Or the social construction of societies in space. In: Space Technology and Applications International Forum (STAIF) Conference Proceedings. Paper Published in Proceedings and Presentation at 2006 STAIF Conference; Albuquerque, NM. Vol. 813. American Institute of Physics (AIP); 2006. pp. 1153-1161. Available from: http://www.astrosociology.org/Library/PDF/submissions/STAIF_Astrosociology%20of%20Space%20ColoniesPDF.pdf

[48] Pass J. Chapter 9: An astrosociological perspective on the societal impact of spaceflight. In: Dick SJ, editor. Historical Studies in the Societal Impact of Spaceflight.

Washington, DC: National Aeronautics and Space Administration (NASA) History Program Office; 2015. p. 554. Available from: http://www.astrosociology.org/Library/PDF/Historical-Studies-Societal-Impact-Spaceflight.pdf

[49] Lee P. Habitability of lava tubes on the Moon and Mars: Lessons from Earth. In: Final Summary to 2018 MBR Space Settlement Challenge. Dubai Future Foundation; 2019. Available from: Guaana.com; https://digital.lib.usf.edu/content/SF/S0/07/02/37/00001/K26-05529-PascalLEE-MBRSSC-Guaana-Final_Summary-20190128-A-Compressed.pdf

[50] Sherwood B. Organizing ourselves: Schema to build the international space architecture community. In: AIAA Space Conference and Exposition; San Jose, CA (AIAA 2006-7471). 2006

[51] Harrison AA. Humanizing outer space: Architecture, habitability, and behavioral health. Acta Astronautica. 2010;**66**(5-6):890-896

[52] Pletser V, Zubrin R, Quinn K. Simulation of Martian EVA at the Mars Society Arctic Research Station. In: 53rd International Astronautical Congress, The World Congress; 10-19 October 2002; Houston, TX. 2002. pp. 1-11

[53] Durão MJ. Embodied space: A sensorial approach to spatial experience. In: Proceedings at the First Symposium of Astrosociology That Was Part of the Space Propulsion, and Energy Sciences International Forum Held in Huntsville, Alabama. The Session Was Called "Space Societies: The Settlement of Space Environments". 2009. p. 399. Available from: http://www.astrosociology.org/Library/PDF/Durao_SPESIF2009.pdf

[54] Durkheim E. The Division of Labor in Society (Translated by G. Simpson). New York: The Free Press; 1964

[55] Rainer E. Chapter 5: Projecting landscapes of the human mind onto another world: Changing faces of an imaginary Mars. In: Geppert ACT, editor. Imagining Outer Space: European Astroculture in the Twentieth Century. London: Palgrave Macmillan; 2012. pp. 89-105

[56] Bierstedt R. The Social Order. 3rd ed. New York: McGraw-Hill, Inc.; 1970

[57] Pass J. Deviance in space habitats: A preliminary look at health and safety violations. *Physics Procedia*. 2011;**20**:353-368. Paper Published in Proceedings at the 3rd Symposium of Astrosociology That Was Part of the Space Propulsion, and Energy Sciences International Forum (SPESIF) Held at the University of Maryland, College Park. The Session Was Called "Medical Astrosociology." Available from: http://www.astrosociology.org/Library/PDF/SPESIF2011--Deviance-in-Space-Habitats.pdf

[58] Ambrosius JD. Our cosmic future? How religion might shape it. In: Levinson P, Waltemathe M, editors. Touching the Face of the Cosmos: On the Interaction of Space Travel and Religion. New York: Fordham University Press. p. 28

[59] Pass J. Medical astrosociology: Ethical dilemmas in space environments. In: Paper Presented as Part of an Astrosociology Session at the AIAA Space 2009 Conference; 20 September 2009; Pasadena, CA. 2009. Available from: http://www.astrosociology.org/Library/PDF/Pass_EthicalDilemmas.pdf

[60] Stoner I. Humans should not colonize Mars. Journal of the American Philosophical Association. 2017;**3**(3):334-353

[61] Slobodian RE. Selling space colonization and immorality: A psychosocial, anthropological critique

of the rush to colonize Mars. Acta Astronautica. 2015;**113**:89-104

[62] Marino L. Humanity is not prepared to colonize Mars. Futures. 2019;**110**:15-18

[63] Pass J. Chapter 9: Astrosociology: Social problems on Earth and in outer space. In: Treviño AJ, editor. The Cambridge Handbook of Social Problems. Vol. 1. New York: Cambridge University Press; 2018. pp. 149-166

[64] Pool SL, Davis JR. Space medicine roots: A historical perspective for the current direction. Aviation, Space, and Environmental Medicine. 2007;**78**(Supplement 1):A3-A4

[65] NASA's Twins Study Results Published in Science Journal. 2019. Available from: https://www.nasa.gov/feature/nasa-s-twins-study-results-published-in-science/

[66] Pass J. Space medicine: Medical astrosociology in the sickbay. In: Paper Presented as Part of a Session Entitled "Astrosociological Perspective on Space Exploration" at the AIAA Aerospace Science Meeting and Exhibit (ASM 2008) Conference; 8 January 2008; Reno, NV. 2008. Available from: http://www.astrosociology.org/Library/PDF/ASM2008_MedicalAstrosociology.pdf

[67] NASA Human Research Program. Available from: https://www.nasa.gov/hrp/about

[68] Sipes WE, Polk JD, Beven G, Shepanek M. Behavioral health and performance. In: Nicogossian A, Williams R, Huntoon C, Doarn C, Polk J, Schneider V, editors. Space Physiology and Medicine. New York: Springer; 2016. p. 2016

[69] Miquel J, Economos AC. Space gerontology. NASA conference publication 2248. In: Proceedings Held at NASA Ames Research Center; 30-31 January 1978; Moffett Field, CA. Washington, DC: NASA

[70] Hearsey CM. The nexus between law and astrosociology. Astropolitics. 2011;**9**(1):3

[71] Harrison AA. Overcoming the image of little green men: Astrosociology and SETI. In: Paper Presented as Part of a Dedicated Session Entitled "Astrosociology: The Sociology of Outer Space," at the California Sociological Association (CSA) Conference; 11 November 2005; Sacramento, CA. 2005 Available from: http://www.astrosociology.org/Library/PDF/submissions/Overcoming%20LGM_Harrison.pdf

[72] Montgomery L. Opportunities of first impression and the outer space treaty. Astrosociological Insights. 2017;**6**(1):8-10

[73] NASA, Office of Safety and Mission Assurance (OSMA). Available from: https://sma.nasa.gov/sma-disciplines/planetary-protection

[74] Harrison AA, Connell K, editors. Workshop on the Societal Implications of Astrobiology: Final Report, Ames Research Center: NASA Technical Memorandum. 2001. Available from: http://astrobiology.arc.nasa.gov/workshops/societal/societal_report.pdf; http://www.astrosociology.org/Library/PDF/NASA-Workshop-Report-Societal-Implications-of-Astrobiology.pdf

[75] Dick SJ. Astrobiology, Discovery, and Societal Impact. Cambridge, England: Cambridge University Press; 2018. p. 174

[76] Dick SJ. Analogy and the societal implications of astrobiology. Astropolitics. 2014;**12**(2-3):210-230

[77] Pass J. Exo-astrosociology and the search for technosignatures. In: Paper Presented as Part of the Society

and Aerospace Technology Session at the AIAA SciTech 2019 Conference; 9 January 2019; San Diego, CA. 2019. Available from: http://www.astrosociology.org/Library/PDF/JPass--Exo-AstrosociologyAndTechnosignatures.pdf

[78] Pass J. Astrosociology implications of astrobiology (revisited). In: Proceedings at the 2nd Symposium on Astrosociology that Was Part of the Space Proposition, and Energy International Forum (SPESIF) Held at the Johns Hopkins University Applied Physics Laboratory, Laurel, Maryland. The session was called Astrosociology and Astrobiology (and SETI). American Institute of Physics (AIP) Proceedings. Vol. 1208. 2010. pp. 402-417. Available from: http://astrosociology.org/Library/PDF/SPESIF2010_Pass_Astrobiology.pdf

[79] NASA Spacecraft Detects Buried Glaciers on Mars. 2008. Available from: https://www.nasa.gov/mission_pages/MRO/news/mro-20081120.html#:~:text=%2D%20NASA's%20Mars%20Reconnaissance%20Orbiter%20has,identified%20on%20the%20Red%20Planet.&text=This%20discovery%20is%20similar%20to,under%20rocky%20coverings%20in%20Antarctica

[80] Capelotti PJ. Space: The [archaeological] frontier. Archaeology. 2004;**57**(6):46-51

[81] Gorman A. We Need to Protect the Heritage of the Apollo Missions. 2009. Available from: theconversation.com; https://theconversation.com/we-need-to-protect-the-heritage-of-the-apollo-missions-117007

[82] B612. Available from: https://b612foundation.org/

[83] The Planetary Society Planetary Defense. Available from: https://www.planetary.org/explore/projects/planetary-defense/

[84] Planetary Defense. The National Space Society. Available from: https://space.nss.org/planetary-defense/

[85] Smith M. NASA and the Space Force to Work Together on Planetary Defense, Space Policy Online.com. 2020. Available from: https://spacepolicyonline.com/news/nasa-and-space-force-to-work-together-on-planetary-defense/

[86] NASA and the Space Force to Work Together on Planetary Defense. Place Policy Online. 2020. Available from: https://spacepolicyonline.com/news/nasa-and-space-force-to-work-together-on-planetary-defense/#:~:text=NASA%20and%20the%20new%20U.S.,comets%20%E2%80%94%20that%20may%20threaten%20Earth.&text=A%20collision%20with%20a%20NEO,are%20likely%20to%20impact%20Earth

[87] ESA Safety and Security Planetary Defence. Available from: https://www.esa.int/Safety_Security/Planetary_Defence

[88] Pass J. Applied astrosociology: The new imperative to protect the Earth and human societies. In: Paper Presented as Part of a Session Entitled "The Astrosociology of Space Colonization" at the AIAA Space 2006 Conference; 20 September 2006; San Jose, CA. 2006. Available from: http://www.astrosociology.org/Library/PDF/Protecting%20Societies.pdf